INTERNATIONAL ENCYCLOPEDIA of UNIFIED SCIENCE

The Structure of Scientific Revolutions

Second Edition, Enlarged

By

Thomas S. Kuhn

VOLUMES I AND II · FOUNDATIONS OF THE UNITY OF SCIENCE

VOLUME II · NUMBER 2

International Encyclopedia of Unified Science

Editor-in-Chief Otto Neurath†
Associate Editors Rudolf Carnap Charles Morris

Foundations of the Unity of Science

(Volumes I–II of the Encyclopedia)

SBN: 226–45803–2 (clothbound); 226–45804–0 (paperback)
Library of Congress Catalog Card Number: 79–107472

The University of Chicago Press, Chicago 60637
The University of Chicago Press, Ltd., London

International Encyclopedia of Unified Science

Volume 2 · Number 2

The Structure of Scientific Revolutions

Thomas S. Kuhn

Contents:

Preface

The essay that follows is the first full published report on a project originally conceived almost fifteen years ago. At that time I was a graduate student in theoretical physics already within sight of the end of my dissertation. A fortunate involvement with an experimental college course treating physical science for the non-scientist provided my first exposure to the history of science. To my complete surprise, that exposure to out-of-date scientific theory and practice radically undermined some of my basic conceptions about the nature of science and the reasons for its special success.

Those conceptions were ones I had previously drawn partly from scientific training itself and partly from a long-standing avocational interest in the philosophy of science. Somehow, whatever their pedagogic utility and their abstract plausibility, those notions did not at all fit the enterprise that historical study displayed. Yet they were and are fundamental to many discussions of science, and their failures of verisimilitude therefore seemed thoroughly worth pursuing. The result was a drastic shift in my career plans, a shift from physics to history of science and then, gradually, from relatively straightforward historical problems back to the more philosophical concerns that had initially led me to history. Except for a few articles, this essay is the first of my published works in which these early concerns are dominant. In some part it is an attempt to explain to myself and to friends how I happened to be drawn from science to its history in the first place.

My first opportunity to pursue in depth some of the ideas set forth below was provided by three years as a Junior Fellow of the Society of Fellows of Harvard University. Without that period of freedom the transition to a new field of study would have been far more difficult and might not have been achieved. Part of my time in those years was devoted to history of science proper. In particular I continued to study the writings of Alexandre Koyré and first encountered those of Emile Meyerson, Hélène Metzger, and Anneliese Maier.[1] More clearly than most other recent scholars, this group has shown what it was like to think scientifically in a period when the canons of scientific thought were very different from those current today. Though I increasingly question a few of their particular historical interpretations, their works, together with A. O. Lovejoy's *Great Chain of Being*, have been second only to primary source materials in shaping my conception of what the history of scientific ideas can be.

Much of my time in those years, however, was spent exploring fields without apparent relation to history of science but in which research now discloses problems like the ones history was bringing to my attention. A footnote encountered by chance led me to the experiments by which Jean Piaget has illuminated both the various worlds of the growing child and the process of transition from one to the next.[2] One of my colleagues set me to reading papers in the psychology of perception, particularly the Gestalt psychologists; another introduced me to B. L. Whorf's speculations about the effect of language on world view; and W. V. O. Quine opened for me the philosophical puzzles of the analytic-synthetic distinction.[3] That is the sort of random exploration that the Society of Fellows permits, and only through it could I have encountered Ludwik Fleck's almost unknown monograph, *Entstehung und Entwicklung einer wis-*

[1] Particularly influential were Alexandre Koyré, *Etudes Galiléennes* (3 vols.; Paris, 1939); Emile Meyerson, *Identity and Reality*, trans. Kate Loewenberg (New York, 1930); Hélène Metzger, *Les doctrines chimiques en France du début du XVIIe à la fin du XVIIIe siècle* (Paris, 1923), and *Newton, Stahl, Boerhaave et la doctrine chimique* (Paris, 1930); and Anneliese Maier, *Die Vorläufer Galileis im 14. Jahrhundert* ("Studien zur Naturphilosophie der Spätscholastik"; Rome, 1949).

[2] Because they displayed concepts and processes that also emerge directly from the history of science, two sets of Piaget's investigations proved particularly important: *The Child's Conception of Causality*, trans. Marjorie Gabain (London, 1930), and *Les notions de mouvement et de vitesse chez l'enfant* (Paris, 1946).

[3] Whorf's papers have since been collected by John B. Carroll, *Language, Thought, and Reality—Selected Writings of Benjamin Lee Whorf* (New York, 1956). Quine has presented his views in "Two Dogmas of Empiricism," reprinted in his *From a Logical Point of View* (Cambridge, Mass., 1953), pp. 20–46.

senschaftlichen Tatsache (Basel, 1935), an essay that anticipates many of my own ideas. Together with a remark from another Junior Fellow, Francis X. Sutton, Fleck's work made me realize that those ideas might require to be set in the sociology of the scientific community. Though readers will find few references to either these works or conversations below, I am indebted to them in more ways than I can now reconstruct or evaluate.

During my last year as a Junior Fellow, an invitation to lecture for the Lowell Institute in Boston provided a first chance to try out my still developing notion of science. The result was a series of eight public lectures, delivered during March, 1951, on "The Quest for Physical Theory." In the next year I began to teach history of science proper, and for almost a decade the problems of instructing in a field I had never systematically studied left little time for explicit articulation of the ideas that had first brought me to it. Fortunately, however, those ideas proved a source of implicit orientation and of some problem-structure for much of my more advanced teaching. I therefore have my students to thank for invaluable lessons both about the viability of my views and about the techniques appropriate to their effective communication. The same problems and orientation give unity to most of the dominantly historical, and apparently diverse, studies I have published since the end of my fellowship. Several of them deal with the integral part played by one or another metaphysic in creative scientific research. Others examine the way in which the experimental bases of a new theory are accumulated and assimilated by men committed to an incompatible older theory. In the process they describe the type of development that I have below called the "emergence" of a new theory or discovery. There are other such ties besides.

The final stage in the development of this essay began with an invitation to spend the year 1958–59 at the Center for Advanced Studies in the Behavioral Sciences. Once again I was able to give undivided attention to the problems discussed below. Even more important, spending the year in a community composed predominantly of social scientists confronted me with unanticipated problems about the differences between such communities and those of the natural scientists among whom I had been trained. Particularly, I was struck by the number and extent of the overt disagreements between social scientists about the nature of legitimate scientific problems and methods. Both history and acquaintance made me doubt that practitioners of the natural sciences possess firmer or more permanent answers to such questions than their colleagues in social science. Yet, somehow, the practice of astronomy, physics, chemistry, or biology normally fails to evoke the controversies over fundamentals that today often seem endemic among, say, psychologists or sociologists. Attempting to discover the source of that difference led me to recognize the role in scientific research of what I have since called "paradigms." These I take to be universally recognized scientific achievements that for a time provide model problems and solutions to a community of practitioners. Once that piece of my puzzle fell into place, a draft of this essay emerged rapidly.

The subsequent history of that draft need not be recounted here, but a few words must be said about the form that it has preserved through revisions. Until a first version had been completed and largely revised, I anticipated that the manuscript would appear exclusively as a volume in the *Encyclopedia of Unified Science*. The editors of that pioneering work had first solicited it, then held me firmly to a commitment, and finally waited with extraordinary tact and patience for a result. I am much indebted to them, particularly to Charles Morris, for wielding the essential goad and for advising me about the manuscript that resulted. Space limits of the *Encyclopedia* made it necessary, however, to present my views in an extremely condensed and schematic form. Though subsequent events have somewhat relaxed those restrictions and have made possible simultaneous independent publication, this work remains an essay rather than the full-scale book my subject will ultimately demand.

Since my most fundamental objective is to urge a change in

the perception and evaluation of familiar data, the schematic character of this first presentation need be no drawback. On the contrary, readers whose own research has prepared them for the sort of reorientation here advocated may find the essay form both more suggestive and easier to assimilate. But it has disadvantages as well, and these may justify my illustrating at the very start the sorts of extension in both scope and depth that I hope ultimately to include in a longer version. Far more historical evidence is available than I have had space to exploit below. Furthermore, that evidence comes from the history of biological as well as of physical science. My decision to deal here exclusively with the latter was made partly to increase this essay's coherence and partly on grounds of present competence. In addition, the view of science to be developed here suggests the potential fruitfulness of a number of new sorts of research, both historical and sociological. For example, the manner in which anomalies, or violations of expectation, attract the increasing attention of a scientific community needs detailed study, as does the emergence of the crises that may be induced by repeated failure to make an anomaly conform. Or again, if I am right that each scientific revolution alters the historical perspective of the community that experiences it, then that change of perspective should affect the structure of postrevolutionary textbooks and research publications. One such effect—a shift in the distribution of the technical literature cited in the footnotes to research reports—ought to be studied as a possible index to the occurrence of revolutions.

The need for drastic condensation has also forced me to forego discussion of a number of major problems. My distinction between the pre- and the post-paradigm periods in the development of a science is, for example, much too schematic. Each of the schools whose competition characterizes the earlier period is guided by something much like a paradigm; there are circumstances, though I think them rare, under which two paradigms can coexist peacefully in the later period. Mere possession of a paradigm is not quite a sufficient criterion for the developmental transition discussed in Section II. More important, except in occasional brief asides, I have said nothing about the role of technological advance or of external social, economic, and intellectual conditions in the development of the sciences. One need, however, look no further than Copernicus and the calendar to discover that external conditions may help to transform a mere anomaly into a source of acute crisis. The same example would illustrate the way in which conditions outside the sciences may influence the range of alternatives available to the man who seeks to end a crisis by proposing one or another revolutionary reform.[4] Explicit consideration of effects like these would not, I think, modify the main theses developed in this essay, but it would surely add an analytic dimension of first-rate importance for the understanding of scientific advance.

Finally, and perhaps most important of all, limitations of space have drastically affected my treatment of the philosophical implications of this essay's historically oriented view of science. Clearly, there are such implications, and I have tried both to point out and to document the main ones. But in doing so I have usually refrained from detailed discussion of the various positions taken by contemporary philosophers on the corresponding issues. Where I have indicated skepticism, it has more often been directed to a philosophical attitude than to any one of its fully articulated expressions. As a result, some of those who know and work within one of those articulated positions may feel that I have missed their point. I think they will be wrong, but this essay is not calculated to convince them. To attempt that would have required a far longer and very different sort of book.

The autobiographical fragments with which this preface

[4] These factors are discussed in T. S. Kuhn, *The Copernican Revolution: Planetary Astronomy in the Development of Western Thought* (Cambridge, Mass., 1957), pp. 122–32, 270–71. Other effects of external intellectual and economic conditions upon substantive scientific development are illustrated in my papers, "Conservation of Energy as an Example of Simultaneous Discovery," *Critical Problems in the History of Science*, ed. Marshall Clagett (Madison, Wis., 1959), pp. 321–56; "Engineering Precedent for the Work of Sadi Carnot," *Archives internationales d'histoire des sciences*, XIII (1960), 247–51; and "Sadi Carnot and the Cagnard Engine," *Isis*, LII (1961), 567–74. It is, therefore, only with respect to the problems discussed in this essay that I take the role of external factors to be minor.

opens will serve to acknowledge what I can recognize of my main debt both to the works of scholarship and to the institutions that have helped give form to my thought. The remainder of that debt I shall try to discharge by citation in the pages that follow. Nothing said above or below, however, will more than hint at the number and nature of my personal obligations to the many individuals whose suggestions and criticisms have at one time or another sustained and directed my intellectual development. Too much time has elapsed since the ideas in this essay began to take shape; a list of all those who may properly find some signs of their influence in its pages would be almost coextensive with a list of my friends and acquaintances. Under the circumstances, I must restrict myself to the few most significant influences that even a faulty memory will never entirely suppress.

It was James B. Conant, then president of Harvard University, who first introduced me to the history of science and thus initiated the transformation in my conception of the nature of scientific advance. Ever since that process began, he has been generous of his ideas, criticisms, and time—including the time required to read and suggest important changes in the draft of my manuscript. Leonard K. Nash, with whom for five years I taught the historically oriented course that Dr. Conant had started, was an even more active collaborator during the years when my ideas first began to take shape, and he has been much missed during the later stages of their development. Fortunately, however, after my departure from Cambridge, his place as creative sounding board and more was assumed by my Berkeley colleague, Stanley Cavell. That Cavell, a philosopher mainly concerned with ethics and aesthetics, should have reached conclusions quite so congruent to my own has been a constant source of stimulation and encouragement to me. He is, furthermore, the only person with whom I have ever been able to explore my ideas in incomplete sentences. That mode of communication attests an understanding that has enabled him to point me the way through or around several major barriers encountered while preparing my first manuscript.

Since that version was drafted, many other friends have helped with its reformulation. They will, I think, forgive me if I name only the four whose contributions proved most far-reaching and decisive: Paul K. Feyerabend of Berkeley, Ernest Nagel of Columbia, H. Pierre Noyes of the Lawrence Radiation Laboratory, and my student, John L. Heilbron, who has often worked closely with me in preparing a final version for the press. I have found all their reservations and suggestions extremely helpful, but I have no reason to believe (and some reason to doubt) that either they or the others mentioned above approve in its entirety the manuscript that results.

My final acknowledgments, to my parents, wife, and children, must be of a rather different sort. In ways which I shall probably be the last to recognize, each of them, too, has contributed intellectual ingredients to my work. But they have also, in varying degrees, done something more important. They have, that is, let it go on and even encouraged my devotion to it. Anyone who has wrestled with a project like mine will recognize what it has occasionally cost them. I do not know how to give them thanks.

T. S. K.

BERKELEY, CALIFORNIA
February 1962

I. Introduction: A Role for History

History, if viewed as a repository for more than anecdote or chronology, could produce a decisive transformation in the image of science by which we are now possessed. That image has previously been drawn, even by scientists themselves, mainly from the study of finished scientific achievements as these are recorded in the classics and, more recently, in the textbooks from which each new scientific generation learns to practice its trade. Inevitably, however, the aim of such books is persuasive and pedagogic; a concept of science drawn from them is no more likely to fit the enterprise that produced them than an image of a national culture drawn from a tourist brochure or a language text. This essay attempts to show that we have been misled by them in fundamental ways. Its aim is a sketch of the quite different concept of science that can emerge from the historical record of the research activity itself.

Even from history, however, that new concept will not be forthcoming if historical data continue to be sought and scrutinized mainly to answer questions posed by the unhistorical stereotype drawn from science texts. Those texts have, for example, often seemed to imply that the content of science is uniquely exemplified by the observations, laws, and theories described in their pages. Almost as regularly, the same books have been read as saying that scientific methods are simply the ones illustrated by the manipulative techniques used in gathering textbook data, together with the logical operations employed when relating those data to the textbook's theoretical generalizations. The result has been a concept of science with profound implications about its nature and development.

If science is the constellation of facts, theories, and methods collected in current texts, then scientists are the men who, successfully or not, have striven to contribute one or another element to that particular constellation. Scientific development becomes the piecemeal process by which these items have been added, singly and in combination, to the ever growing stockpile that constitutes scientific technique and knowledge. And history of science becomes the discipline that chronicles both these successive increments and the obstacles that have inhibited their accumulation. Concerned with scientific development, the historian then appears to have two main tasks. On the one hand, he must determine by what man and at what point in time each contemporary scientific fact, law, and theory was discovered or invented. On the other, he must describe and explain the congeries of error, myth, and superstition that have inhibited the more rapid accumulation of the constituents of the modern science text. Much research has been directed to these ends, and some still is.

In recent years, however, a few historians of science have been finding it more and more difficult to fulfil the functions that the concept of development-by-accumulation assigns to them. As chroniclers of an incremental process, they discover that additional research makes it harder, not easier, to answer questions like: When was oxygen discovered? Who first conceived of energy conservation? Increasingly, a few of them suspect that these are simply the wrong sorts of questions to ask. Perhaps science does not develop by the accumulation of individual discoveries and inventions. Simultaneously, these same historians confront growing difficulties in distinguishing the "scientific" component of past observation and belief from what their predecessors had readily labeled "error" and "superstition." The more carefully they study, say, Aristotelian dynamics, phlogistic chemistry, or caloric thermodynamics, the more certain they feel that those once current views of nature were, as a whole, neither less scientific nor more the product of human idiosyncrasy than those current today. If these out-of-date beliefs are to be called myths, then myths can be produced by the same sorts of methods and held for the same sorts of reasons that now lead to scientific knowledge. If, on the other hand, they are to be called science, then science has included bodies of belief quite incompatible with the ones we hold today. Given these alternatives, the historian must choose the latter. Out-of-date theories are not in principle unscientific because they have been discarded. That choice, however, makes it difficult to see scientific development as a process of accretion. The same historical research that displays the difficulties in isolating individual inventions and discoveries gives ground for profound doubts about the cumulative process through which these individual contributions to science were thought to have been compounded.

The result of all these doubts and difficulties is a historiographic revolution in the study of science, though one that is still in its early stages. Gradually, and often without entirely realizing they are doing so, historians of science have begun to ask new sorts of questions and to trace different, and often less than cumulative, developmental lines for the sciences. Rather than seeking the permanent contributions of an older science to our present vantage, they attempt to display the historical integrity of that science in its own time. They ask, for example, not about the relation of Galileo's views to those of modern science, but rather about the relationship between his views and those of his group, i.e., his teachers, contemporaries, and immediate successors in the sciences. Furthermore, they insist upon studying the opinions of that group and other similar ones from the viewpoint—usually very different from that of modern science—that gives those opinions the maximum internal coherence and the closest possible fit to nature. Seen through the works that result, works perhaps best exemplified in the writings of Alexandre Koyré, science does not seem altogether the same enterprise as the one discussed by writers in the older historiographic tradition. By implication, at least, these historical studies suggest the possibility of a new image of science. This essay aims to delineate that image by making explicit some of the new historiography's implications.

What aspects of science will emerge to prominence in the course of this effort? First, at least in order of presentation, is the insufficiency of methodological directives, by themselves, to dictate a unique substantive conclusion to many sorts of scientific questions. Instructed to examine electrical or chemical phenomena, the man who is ignorant of these fields but who knows what it is to be scientific may legitimately reach any one of a number of incompatible conclusions. Among those legitimate possibilities, the particular conclusions he does arrive at are probably determined by his prior experience in other fields, by the accidents of his investigation, and by his own individual makeup. What beliefs about the stars, for example, does he bring to the study of chemistry or electricity? Which of the many conceivable experiments relevant to the new field does he elect to perform first? And what aspects of the complex phenomenon that then results strike him as particularly relevant to an elucidation of the nature of chemical change or of electrical affinity? For the individual, at least, and sometimes for the scientific community as well, answers to questions like these are often essential determinants of scientific development. We shall note, for example, in Section II that the early developmental stages of most sciences have been characterized by continual competition between a number of distinct views of nature, each partially derived from, and all roughly compatible with, the dictates of scientific observation and method. What differentiated these various schools was not one or another failure of method—they were all "scientific"—but what we shall come to call their incommensurable ways of seeing the world and of practicing science in it. Observation and experience can and must drastically restrict the range of admissible scientific belief, else there would be no science. But they cannot alone determine a particular body of such belief. An apparently arbitrary element, compounded of personal and historical accident, is always a formative ingredient of the beliefs espoused by a given scientific community at a given time.

That element of arbitrariness does not, however, indicate that any scientific group could practice its trade without some set of received beliefs. Nor does it make less consequential the particular constellation to which the group, at a given time, is in fact committed. Effective research scarcely begins before a scientific community thinks it has acquired firm answers to questions like the following: What are the fundamental entities of which the universe is composed? How do these interact with each other and with the senses? What questions may legitimately be asked about such entities and what techniques employed in seeking solutions? At least in the mature sciences, answers (or full substitutes for answers) to questions like these are firmly embedded in the educational initiation that prepares and licenses the student for professional practice. Because that education is both rigorous and rigid, these answers come to exert a deep hold on the scientific mind. That they can do so does much to account both for the peculiar efficiency of the normal research activity and for the direction in which it proceeds at any given time. When examining normal science in Sections III, IV, and V, we shall want finally to describe that research as a strenuous and devoted attempt to force nature into the conceptual boxes supplied by professional education. Simultaneously, we shall wonder whether research could proceed without such boxes, whatever the element of arbitrariness in their historic origins and, occasionally, in their subsequent development.

Yet that element of arbitrariness is present, and it too has an important effect on scientific development, one which will be examined in detail in Sections VI, VII, and VIII. Normal science, the activity in which most scientists inevitably spend almost all their time, is predicated on the assumption that the scientific community knows what the world is like. Much of the success of the enterprise derives from the community's willingness to defend that assumption, if necessary at considerable cost. Normal science, for example, often suppresses fundamental novelties because they are necessarily subversive of its basic commitments. Nevertheless, so long as those commitments retain an element of the arbitrary, the very nature of normal research ensures that novelty shall not be suppressed for very long. Sometimes a normal problem, one that ought to be solvable by known rules and procedures, resists the reiterated onslaught of the ablest members of the group within whose competence it falls. On other occasions a piece of equipment designed and constructed for the purpose of normal research fails to perform in the anticipated manner, revealing an anomaly that cannot, despite repeated effort, be aligned with professional expectation. In these and other ways besides, normal science repeatedly goes astray. And when it does—when, that is, the profession can no longer evade anomalies that subvert the existing tradition of scientific practice—then begin the extraordinary investigations that lead the profession at last to a new set of commitments, a new basis for the practice of science. The extraordinary episodes in which that shift of professional commitments occurs are the ones known in this essay as scientific revolutions. They are the tradition-shattering complements to the tradition-bound activity of normal science.

The most obvious examples of scientific revolutions are those famous episodes in scientific development that have often been labeled revolutions before. Therefore, in Sections IX and X, where the nature of scientific revolutions is first directly scrutinized, we shall deal repeatedly with the major turning points in scientific development associated with the names of Copernicus, Newton, Lavoisier, and Einstein. More clearly than most other episodes in the history of at least the physical sciences, these display what all scientific revolutions are about. Each of them necessitated the community's rejection of one time-honored scientific theory in favor of another incompatible with it. Each produced a consequent shift in the problems available for scientific scrutiny and in the standards by which the profession determined what should count as an admissible problem or as a legitimate problem-solution. And each transformed the scientific imagination in ways that we shall ultimately need to describe as a transformation of the world within which scientific work was done. Such changes, together with the controversies that almost always accompany them, are the defining characteristics of scientific revolutions.

These characteristics emerge with particular clarity from a study of, say, the Newtonian or the chemical revolution. It is, however, a fundamental thesis of this essay that they can also be retrieved from the study of many other episodes that were not so obviously revolutionary. For the far smaller professional group affected by them, Maxwell's equations were as revolutionary as Einstein's, and they were resisted accordingly. The invention of other new theories regularly, and appropriately, evokes the same response from some of the specialists on whose area of special competence they impinge. For these men the new theory implies a change in the rules governing the prior practice of normal science. Inevitably, therefore, it reflects upon much scientific work they have already successfully completed. That is why a new theory, however special its range of application, is seldom or never just an increment to what is already known. Its assimilation requires the reconstruction of prior theory and the re-evaluation of prior fact, an intrinsically revolutionary process that is seldom completed by a single man and never overnight. No wonder historians have had difficulty in dating precisely this extended process that their vocabulary impels them to view as an isolated event.

Nor are new inventions of theory the only scientific events that have revolutionary impact upon the specialists in whose domain they occur. The commitments that govern normal science specify not only what sorts of entities the universe does contain, but also, by implication, those that it does not. It follows, though the point will require extended discussion, that a discovery like that of oxygen or X-rays does not simply add one more item to the population of the scientist's world. Ultimately it has that effect, but not until the professional community has re-evaluated traditional experimental procedures, altered its conception of entities with which it has long been familiar, and, in the process, shifted the network of theory through which it deals with the world. Scientific fact and theory are not categorically separable, except perhaps within a single tradition of normal-scientific practice. That is why the unexpected discovery is not simply factual in its import and why the scientist's world is qualitatively transformed as well as quantitatively enriched by fundamental novelties of either fact or theory.

This extended conception of the nature of scientific revolutions is the one delineated in the pages that follow. Admittedly the extension strains customary usage. Nevertheless, I shall continue to speak even of discoveries as revolutionary, because it is just the possibility of relating their structure to that of, say, the Copernican revolution that makes the extended conception seem to me so important. The preceding discussion indicates how the complementary notions of normal science and of scientific revolutions will be developed in the nine sections immediately to follow. The rest of the essay attempts to dispose of three remaining central questions. Section XI, by discussing the textbook tradition, considers why scientific revolutions have previously been so difficult to see. Section XII describes the revolutionary competition between the proponents of the old normal-scientific tradition and the adherents of the new one. It thus considers the process that should somehow, in a theory of scientific inquiry, replace the confirmation or falsification procedures made familiar by our usual image of science. Competition between segments of the scientific community is the only historical process that ever actually results in the rejection of one previously accepted theory or in the adoption of another. Finally, Section XIII will ask how development through revolutions can be compatible with the apparently unique character of scientific progress. For that question, however, this essay will provide no more than the main outlines of an answer, one which depends upon characteristics of the scientific community that require much additional exploration and study.

Undoubtedly, some readers will already have wondered whether historical study can possibly effect the sort of conceptual transformation aimed at here. An entire arsenal of dichotomies is available to suggest that it cannot properly do so. History, we too often say, is a purely descriptive discipline. The theses suggested above are, however, often interpretive and sometimes normative. Again, many of my generalizations are about the sociology or social psychology of scientists; yet at least a few of my conclusions belong traditionally to logic or epistemology. In the preceding paragraph I may even seem to have violated the very influential contemporary distinction between "the context of discovery" and "the context of justification." Can anything more than profound confusion be indicated by this admixture of diverse fields and concerns?

Having been weaned intellectually on these distinctions and others like them, I could scarcely be more aware of their import and force. For many years I took them to be about the nature of knowledge, and I still suppose that, appropriately recast, they have something important to tell us. Yet my attempts to apply them, even *grosso modo,* to the actual situations in which knowledge is gained, accepted, and assimilated have made them seem extraordinarily problematic. Rather than being elementary logical or methodological distinctions, which would thus be prior to the analysis of scientific knowledge, they now seem integral parts of a traditional set of substantive answers to the very questions upon which they have been deployed. That circularity does not at all invalidate them. But it does make them parts of a theory and, by doing so, subjects them to the same scrutiny regularly applied to theories in other fields. If they are to have more than pure abstraction as their content, then that content must be discovered by observing them in application to the data they are meant to elucidate. How could history of science fail to be a source of phenomena to which theories about knowledge may legitimately be asked to apply?

II. The Route to Normal Science

In this essay, 'normal science' means research firmly based upon one or more past scientific achievements, achievements that some particular scientific community acknowledges for a time as supplying the foundation for its further practice. Today such achievements are recounted, though seldom in their original form, by science textbooks, elementary and advanced. These textbooks expound the body of accepted theory, illustrate many or all of its successful applications, and compare these applications with exemplary observations and experiments. Before such books became popular early in the nineteenth century (and until even more recently in the newly matured sciences), many of the famous classics of science fulfilled a similar function. Aristotle's *Physica,* Ptolemy's *Almagest,* Newton's *Principia* and *Opticks,* Franklin's *Electricity,* Lavoisier's *Chemistry,* and Lyell's *Geology*—these and many other works served for a time implicitly to define the legitimate problems and methods of a research field for succeeding generations of practitioners. They were able to do so because they shared two essential characteristics. Their achievement was sufficiently unprecedented to attract an enduring group of adherents away from competing modes of scientific activity. Simultaneously, it was sufficiently open-ended to leave all sorts of problems for the redefined group of practitioners to resolve.

Achievements that share these two characteristics I shall henceforth refer to as 'paradigms,' a term that relates closely to 'normal science.' By choosing it, I mean to suggest that some accepted examples of actual scientific practice—examples which include law, theory, application, and instrumentation together—provide models from which spring particular coherent traditions of scientific research. These are the traditions which the historian describes under such rubrics as 'Ptolemaic astronomy' (or 'Copernican'), 'Aristotelian dynamics' (or 'Newtonian'), 'corpuscular optics' (or 'wave optics'), and so on. The study of paradigms, including many that are far more specialized than those named illustratively above, is what mainly prepares the student for membership in the particular scientific community with which he will later practice. Because he there joins men who learned the bases of their field from the same concrete models, his subsequent practice will seldom evoke overt disagreement over fundamentals. Men whose research is based on shared paradigms are committed to the same rules and standards for scientific practice. That commitment and the apparent consensus it produces are prerequisites for normal science, i.e., for the genesis and continuation of a particular research tradition.

Because in this essay the concept of a paradigm will often substitute for a variety of familiar notions, more will need to be said about the reasons for its introduction. Why is the concrete scientific achievement, as a locus of professional commitment, prior to the various concepts, laws, theories, and points of view that may be abstracted from it? In what sense is the shared paradigm a fundamental unit for the student of scientific development, a unit that cannot be fully reduced to logically atomic components which might function in its stead? When we encounter them in Section V, answers to these questions and to others like them will prove basic to an understanding both of normal science and of the associated concept of paradigms. That more abstract discussion will depend, however, upon a previous exposure to examples of normal science or of paradigms in operation. In particular, both these related concepts will be clarified by noting that there can be a sort of scientific research without paradigms, or at least without any so unequivocal and so binding as the ones named above. Acquisition of a paradigm and of the more esoteric type of research it permits is a sign of maturity in the development of any given scientific field.

If the historian traces the scientific knowledge of any selected group of related phenomena backward in time, he is likely to encounter some minor variant of a pattern here illustrated from the history of physical optics. Today's physics textbooks tell the student that light is photons, i.e., quantum-mechanical entities that exhibit some characteristics of waves and some of particles. Research proceeds accordingly, or rather according to the more elaborate and mathematical characterization from which this usual verbalization is derived. That characterization of light is, however, scarcely half a century old. Before it was developed by Planck, Einstein, and others early in this century, physics texts taught that light was transverse wave motion, a conception rooted in a paradigm that derived ultimately from the optical writings of Young and Fresnel in the early nineteenth century. Nor was the wave theory the first to be embraced by almost all practitioners of optical science. During the eighteenth century the paradigm for this field was provided by Newton's *Opticks,* which taught that light was material corpuscles. At that time physicists sought evidence, as the early wave theorists had not, of the pressure exerted by light particles impinging on solid bodies.[1]

These transformations of the paradigms of physical optics are scientific revolutions, and the successive transition from one paradigm to another via revolution is the usual developmental pattern of mature science. It is not, however, the pattern characteristic of the period before Newton's work, and that is the contrast that concerns us here. No period between remote antiquity and the end of the seventeenth century exhibited a single generally accepted view about the nature of light. Instead there were a number of competing schools and subschools, most of them espousing one variant or another of Epicurean, Aristotelian, or Platonic theory. One group took light to be particles emanating from material bodies; for another it was a modification of the medium that intervened between the body and the eye; still another explained light in terms of an interaction of the medium with an emanation from the eye; and there were other combinations and modifications besides. Each of the corresponding schools derived strength from its relation to some particular metaphysic, and each emphasized, as para-

[1] Joseph Priestley, *The History and Present State of Discoveries Relating to Vision, Light, and Colours* (London, 1772), pp. 385–90.

digmatic observations, the particular cluster of optical phenomena that its own theory could do most to explain. Other observations were dealt with by *ad hoc* elaborations, or they remained as outstanding problems for further research.[2]

At various times all these schools made significant contributions to the body of concepts, phenomena, and techniques from which Newton drew the first nearly uniformly accepted paradigm for physical optics. Any definition of the scientist that excludes at least the more creative members of these various schools will exclude their modern successors as well. Those men were scientists. Yet anyone examining a survey of physical optics before Newton may well conclude that, though the field's practitioners were scientists, the net result of their activity was something less than science. Being able to take no common body of belief for granted, each writer on physical optics felt forced to build his field anew from its foundations. In doing so, his choice of supporting observation and experiment was relatively free, for there was no standard set of methods or of phenomena that every optical writer felt forced to employ and explain. Under these circumstances, the dialogue of the resulting books was often directed as much to the members of other schools as it was to nature. That pattern is not unfamiliar in a number of creative fields today, nor is it incompatible with significant discovery and invention. It is not, however, the pattern of development that physical optics acquired after Newton and that other natural sciences make familiar today.

The history of electrical research in the first half of the eighteenth century provides a more concrete and better known example of the way a science develops before it acquires its first universally received paradigm. During that period there were almost as many views about the nature of electricity as there were important electrical experimenters, men like Hauksbee, Gray, Desaguliers, Du Fay, Nollett, Watson, Franklin, and others. All their numerous concepts of electricity had something in common—they were partially derived from one or an-

[2] Vasco Ronchi, *Histoire de la lumière,* trans. Jean Taton (Paris, 1956), chaps. i–iv.

other version of the mechanico-corpuscular philosophy that guided all scientific research of the day. In addition, all were components of real scientific theories, of theories that had been drawn in part from experiment and observation and that partially determined the choice and interpretation of additional problems undertaken in research. Yet though all the experiments were electrical and though most of the experimenters read each other's works, their theories had no more than a family resemblance.[3]

One early group of theories, following seventeenth-century practice, regarded attraction and frictional generation as the fundamental electrical phenomena. This group tended to treat repulsion as a secondary effect due to some sort of mechanical rebounding and also to postpone for as long as possible both discussion and systematic research on Gray's newly discovered effect, electrical conduction. Other "electricians" (the term is their own) took attraction and repulsion to be equally elementary manifestations of electricity and modified their theories and research accordingly. (Actually, this group is remarkably small—even Franklin's theory never quite accounted for the mutual repulsion of two negatively charged bodies.) But they had as much difficulty as the first group in accounting simultaneously for any but the simplest conduction effects. Those effects, however, provided the starting point for still a third group, one which tended to speak of electricity as a "fluid" that could run through conductors rather than as an "effluvium" that emanated from non-conductors. This group, in its turn, had difficulty reconciling its theory with a number of attractive and

[3] Duane Roller and Duane H. D. Roller, *The Development of the Concept of Electric Charge: Electricity from the Greeks to Coulomb* ("Harvard Case Histories in Experimental Science," Case 8; Cambridge, Mass., 1954); and I. B. Cohen, *Franklin and Newton: An Inquiry into Speculative Newtonian Experimental Science and Franklin's Work in Electricity as an Example Thereof* (Philadelphia, 1956), chaps. vii–xii. For some of the analytic detail in the paragraph that follows in the text, I am indebted to a still unpublished paper by my student John L. Heilbron. Pending its publication, a somewhat more extended and more precise account of the emergence of Franklin's paradigm is included in T. S. Kuhn, "The Function of Dogma in Scientific Research," in A. C. Crombie (ed.), "Symposium on the History of Science, University of Oxford, July 9–15, 1961," to be published by Heinemann Educational Books, Ltd.

repulsive effects. Only through the work of Franklin and his immediate successors did a theory arise that could account with something like equal facility for very nearly all these effects and that therefore could and did provide a subsequent generation of "electricians" with a common paradigm for its research.

Excluding those fields, like mathematics and astronomy, in which the first firm paradigms date from prehistory and also those, like biochemistry, that arose by division and recombination of specialties already matured, the situations outlined above are historically typical. Though it involves my continuing to employ the unfortunate simplification that tags an extended historical episode with a single and somewhat arbitrarily chosen name (e.g., Newton or Franklin), I suggest that similar fundamental disagreements characterized, for example, the study of motion before Aristotle and of statics before Archimedes, the study of heat before Black, of chemistry before Boyle and Boerhaave, and of historical geology before Hutton. In parts of biology—the study of heredity, for example—the first universally received paradigms are still more recent; and it remains an open question what parts of social science have yet acquired such paradigms at all. History suggests that the road to a firm research consensus is extraordinarily arduous.

History also suggests, however, some reasons for the difficulties encountered on that road. In the absence of a paradigm or some candidate for paradigm, all of the facts that could possibly pertain to the development of a given science are likely to seem equally relevant. As a result, early fact-gathering is a far more nearly random activity than the one that subsequent scientific development makes familiar. Furthermore, in the absence of a reason for seeking some particular form of more recondite information, early fact-gathering is usually restricted to the wealth of data that lie ready to hand. The resulting pool of facts contains those accessible to casual observation and experiment together with some of the more esoteric data retrievable from established crafts like medicine, calendar making, and metallurgy. Because the crafts are one readily accessible source of facts that could not have been casually discovered, technology has often played a vital role in the emergence of new sciences.

But though this sort of fact-collecting has been essential to the origin of many significant sciences, anyone who examines, for example, Pliny's encyclopedic writings or the Baconian natural histories of the seventeenth century will discover that it produces a morass. One somehow hesitates to call the literature that results scientific. The Baconian "histories" of heat, color, wind, mining, and so on, are filled with information, some of it recondite. But they juxtapose facts that will later prove revealing (e.g., heating by mixture) with others (e.g., the warmth of dung heaps) that will for some time remain too complex to be integrated with theory at all.[4] In addition, since any description must be partial, the typical natural history often omits from its immensely circumstantial accounts just those details that later scientists will find sources of important illumination. Almost none of the early "histories" of electricity, for example, mention that chaff, attracted to a rubbed glass rod, bounces off again. That effect seemed mechanical, not electrical.[5] Moreover, since the casual fact-gatherer seldom possesses the time or the tools to be critical, the natural histories often juxtapose descriptions like the above with others, say, heating by antiperistasis (or by cooling), that we are now quite unable to confirm.[6] Only very occasionally, as in the cases of ancient statics, dynamics, and geometrical optics, do facts collected with so little guidance from pre-established theory speak with sufficient clarity to permit the emergence of a first paradigm.

This is the situation that creates the schools characteristic of the early stages of a science's development. No natural history can be interpreted in the absence of at least some implicit body

[4] Compare the sketch for a natural history of heat in Bacon's *Novum Organum*, Vol. VIII of *The Works of Francis Bacon*, ed. J. Spedding, R. L. Ellis, and D. D. Heath (New York, 1869), pp. 179–203.

[5] Roller and Roller, *op. cit.*, pp. 14, 22, 28, 43. Only after the work recorded in the last of these citations do repulsive effects gain general recognition as unequivocally electrical.

[6] Bacon, *op. cit.*, pp. 235, 337, says, "Water slightly warm is more easily frozen than quite cold." For a partial account of the earlier history of this strange observation, see Marshall Clagett, *Giovanni Marliani and Late Medieval Physics* (New York, 1941), chap. iv.

of intertwined theoretical and methodological belief that permits selection, evaluation, and criticism. If that body of belief is not already implicit in the collection of facts—in which case more than "mere facts" are at hand—it must be externally supplied, perhaps by a current metaphysic, by another science, or by personal and historical accident. No wonder, then, that in the early stages of the development of any science different men confronting the same range of phenomena, but not usually all the same particular phenomena, describe and interpret them in different ways. What is surprising, and perhaps also unique in its degree to the fields we call science, is that such initial divergences should ever largely disappear.

For they do disappear to a very considerable extent and then apparently once and for all. Furthermore, their disappearance is usually caused by the triumph of one of the pre-paradigm schools, which, because of its own characteristic beliefs and preconceptions, emphasized only some special part of the too sizable and inchoate pool of information. Those electricians who thought electricity a fluid and therefore gave particular emphasis to conduction provide an excellent case in point. Led by this belief, which could scarcely cope with the known multiplicity of attractive and repulsive effects, several of them conceived the idea of bottling the electrical fluid. The immediate fruit of their efforts was the Leyden jar, a device which might never have been discovered by a man exploring nature casually or at random, but which was in fact independently developed by at least two investigators in the early 1740's.[7] Almost from the start of his electrical researches, Franklin was particularly concerned to explain that strange and, in the event, particularly revealing piece of special apparatus. His success in doing so provided the most effective of the arguments that made his theory a paradigm, though one that was still unable to account for quite all the known cases of electrical repulsion.[8] To be accepted as a paradigm, a theory must seem better than its competitors, but

[7] Roller and Roller, *op. cit.*, pp. 51–54.

[8] The troublesome case was the mutual repulsion of negatively charged bodies, for which see Cohen, *op. cit.*, pp. 491–94, 531–43.

it need not, and in fact never does, explain all the facts with which it can be confronted.

What the fluid theory of electricity did for the subgroup that held it, the Franklinian paradigm later did for the entire group of electricians. It suggested which experiments would be worth performing and which, because directed to secondary or to overly complex manifestations of electricity, would not. Only the paradigm did the job far more effectively, partly because the end of interschool debate ended the constant reiteration of fundamentals and partly because the confidence that they were on the right track encouraged scientists to undertake more precise, esoteric, and consuming sorts of work.[9] Freed from the concern with any and all electrical phenomena, the united group of electricians could pursue selected phenomena in far more detail, designing much special equipment for the task and employing it more stubbornly and systematically than electricians had ever done before. Both fact collection and theory articulation became highly directed activities. The effectiveness and efficiency of electrical research increased accordingly, providing evidence for a societal version of Francis Bacon's acute methodological dictum: "Truth emerges more readily from error than from confusion."[10]

We shall be examining the nature of this highly directed or paradigm-based research in the next section, but must first note briefly how the emergence of a paradigm affects the structure of the group that practices the field. When, in the development of a natural science, an individual or group first produces a synthesis able to attract most of the next generation's practitioners, the older schools gradually disappear. In part their disappear-

[9] It should be noted that the acceptance of Franklin's theory did not end quite all debate. In 1759 Robert Symmer proposed a two-fluid version of that theory, and for many years thereafter electricians were divided about whether electricity was a single fluid or two. But the debates on this subject only confirm what has been said above about the manner in which a universally recognized achievement unites the profession. Electricians, though they continued divided on this point, rapidly concluded that no experimental tests could distinguish the two versions of the theory and that they were therefore equivalent. After that, both schools could and did exploit all the benefits that the Franklinian theory provided (*ibid.*, pp. 543–46, 548–54).

[10] Bacon, *op. cit.*, p. 210.

ance is caused by their members' conversion to the new paradigm. But there are always some men who cling to one or another of the older views, and they are simply read out of the profession, which thereafter ignores their work. The new paradigm implies a new and more rigid definition of the field. Those unwilling or unable to accommodate their work to it must proceed in isolation or attach themselves to some other group.[11] Historically, they have often simply stayed in the departments of philosophy from which so many of the special sciences have been spawned. As these indications hint, it is sometimes just its reception of a paradigm that transforms a group previously interested merely in the study of nature into a profession or, at least, a discipline. In the sciences (though not in fields like medicine, technology, and law, of which the principal *raison d'être* is an external social need), the formation of specialized journals, the foundation of specialists' societies, and the claim for a special place in the curriculum have usually been associated with a group's first reception of a single paradigm. At least this was the case between the time, a century and a half ago, when the institutional pattern of scientific specialization first developed and the very recent time when the paraphernalia of specialization acquired a prestige of their own.

The more rigid definition of the scientific group has other consequences. When the individual scientist can take a paradigm for granted, he need no longer, in his major works, attempt to build his field anew, starting from first principles and justify-

[11] The history of electricity provides an excellent example which could be duplicated from the careers of Priestley, Kelvin, and others. Franklin reports that Nollet, who at mid-century was the most influential of the Continental electricians, "lived to see himself the last of his Sect, except Mr. B.—his Eleve and immediate Disciple" (Max Farrand [ed.], *Benjamin Franklin's Memoirs* [Berkeley, Calif., 1949], pp. 384–86). More interesting, however, is the endurance of whole schools in increasing isolation from professional science. Consider, for example, the case of astrology, which was once an integral part of astronomy. Or consider the continuation in the late eighteenth and early nineteenth centuries of a previously respected tradition of "romantic" chemistry. This is the tradition discussed by Charles C. Gillispie in "The *Encyclopédie* and the Jacobin Philosophy of Science: A Study in Ideas and Consequences," *Critical Problems in the History of Science,* ed. Marshall Clagett (Madison, Wis., 1959), pp. 255–89; and "The Formation of Lamarck's Evolutionary Theory," *Archives internationales d'histoire des sciences,* XXXVII (1956), 323–38.

ing the use of each concept introduced. That can be left to the writer of textbooks. Given a textbook, however, the creative scientist can begin his research where it leaves off and thus concentrate exclusively upon the subtlest and most esoteric aspects of the natural phenomena that concern his group. And as he does this, his research communiqués will begin to change in ways whose evolution has been too little studied but whose modern end products are obvious to all and oppressive to many. No longer will his researches usually be embodied in books addressed, like Franklin's *Experiments . . . on Electricity* or Darwin's *Origin of Species,* to anyone who might be interested in the subject matter of the field. Instead they will usually appear as brief articles addressed only to professional colleagues, the men whose knowledge of a shared paradigm can be assumed and who prove to be the only ones able to read the papers addressed to them.

Today in the sciences, books are usually either texts or retrospective reflections upon one aspect or another of the scientific life. The scientist who writes one is more likely to find his professional reputation impaired than enhanced. Only in the earlier, pre-paradigm, stages of the development of the various sciences did the book ordinarily possess the same relation to professional achievement that it still retains in other creative fields. And only in those fields that still retain the book, with or without the article, as a vehicle for research communication are the lines of professionalization still so loosely drawn that the layman may hope to follow progress by reading the practitioners' original reports. Both in mathematics and astronomy, research reports had ceased already in antiquity to be intelligible to a generally educated audience. In dynamics, research became similarly esoteric in the later Middle Ages, and it recaptured general intelligibility only briefly during the early seventeenth century when a new paradigm replaced the one that had guided medieval research. Electrical research began to require translation for the layman before the end of the eighteenth century, and most other fields of physical science ceased to be generally accessible in the nineteenth. During the same two centuries similar transitions can be isolated in the various parts of the biological sciences. In parts of the social sciences they may well be occurring today. Although it has become customary, and is surely proper, to deplore the widening gulf that separates the professional scientist from his colleagues in other fields, too little attention is paid to the essential relationship between that gulf and the mechanisms intrinsic to scientific advance.

Ever since prehistoric antiquity one field of study after another has crossed the divide between what the historian might call its prehistory as a science and its history proper. These transitions to maturity have seldom been so sudden or so unequivocal as my necessarily schematic discussion may have implied. But neither have they been historically gradual, coextensive, that is to say, with the entire development of the fields within which they occurred. Writers on electricity during the first four decades of the eighteenth century possessed far more information about electrical phenomena than had their sixteenth-century predecessors. During the half-century after 1740, few new sorts of electrical phenomena were added to their lists. Nevertheless, in important respects, the electrical writings of Cavendish, Coulomb, and Volta in the last third of the eighteenth century seem further removed from those of Gray, Du Fay, and even Franklin than are the writings of these early eighteenth-century electrical discoverers from those of the sixteenth century.[12] Sometime between 1740 and 1780, electricians were for the first time enabled to take the foundations of their field for granted. From that point they pushed on to more concrete and recondite problems, and increasingly they then reported their results in articles addressed to other electricians rather than in books addressed to the learned world at large. As a group they achieved what had been gained by astronomers in antiquity

[12] The post-Franklinian developments include an immense increase in the sensitivity of charge detectors, the first reliable and generally diffused techniques for measuring charge, the evolution of the concept of capacity and its relation to a newly refined notion of electric tension, and the quantification of electrostatic force. On all of these see Roller and Roller, *op. cit.*, pp. 66–81; W. C. Walker, "The Detection and Estimation of Electric Charges in the Eighteenth Century," *Annals of Science*, I (1936), 66–100; and Edmund Hoppe, *Geschichte der Elektrizität* (Leipzig, 1884), Part I, chaps. iii–iv.

and by students of motion in the Middle Ages, of physical optics in the late seventeenth century, and of historical geology in the early nineteenth. They had, that is, achieved a paradigm that proved able to guide the whole group's research. Except with the advantage of hindsight, it is hard to find another criterion that so clearly proclaims a field a science.

III. The Nature of Normal Science

What then is the nature of the more professional and esoteric research that a group's reception of a single paradigm permits? If the paradigm represents work that has been done once and for all, what further problems does it leave the united group to resolve? Those questions will seem even more urgent if we now note one respect in which the terms used so far may be misleading. In its established usage, a paradigm is an accepted model or pattern, and that aspect of its meaning has enabled me, lacking a better word, to appropriate 'paradigm' here. But it will shortly be clear that the sense of 'model' and 'pattern' that permits the appropriation is not quite the one usual in defining 'paradigm.' In grammar, for example, '*amo, amas, amat*' is a paradigm because it displays the pattern to be used in conjugating a large number of other Latin verbs, e.g., in producing '*laudo, laudas, laudat*.' In this standard application, the paradigm functions by permitting the replication of examples any one of which could in principle serve to replace it. In a science, on the other hand, a paradigm is rarely an object for replication. Instead, like an accepted judicial decision in the common law, it is an object for further articulation and specification under new or more stringent conditions.

To see how this can be so, we must recognize how very limited in both scope and precision a paradigm can be at the time of its first appearance. Paradigms gain their status because they are more successful than their competitors in solving a few problems that the group of practitioners has come to recognize as acute. To be more successful is not, however, to be either completely successful with a single problem or notably successful with any large number. The success of a paradigm—whether Aristotle's analysis of motion, Ptolemy's computations of planetary position, Lavoisier's application of the balance, or Maxwell's mathematization of the electromagnetic field—is at the start largely a promise of success discoverable in selected and still incomplete examples. Normal science consists in the actualization of that promise, an actualization achieved by extending the knowledge of those facts that the paradigm displays as particularly revealing, by increasing the extent of the match between those facts and the paradigm's predictions, and by further articulation of the paradigm itself.

Few people who are not actually practitioners of a mature science realize how much mop-up work of this sort a paradigm leaves to be done or quite how fascinating such work can prove in the execution. And these points need to be understood. Mopping-up operations are what engage most scientists throughout their careers. They constitute what I am here calling normal science. Closely examined, whether historically or in the contemporary laboratory, that enterprise seems an attempt to force nature into the preformed and relatively inflexible box that the paradigm supplies. No part of the aim of normal science is to call forth new sorts of phenomena; indeed those that will not fit the box are often not seen at all. Nor do scientists normally aim to invent new theories, and they are often intolerant of those invented by others.[1] Instead, normal-scientific research is directed to the articulation of those phenomena and theories that the paradigm already supplies.

Perhaps these are defects. The areas investigated by normal science are, of course, minuscule; the enterprise now under discussion has drastically restricted vision. But those restrictions, born from confidence in a paradigm, turn out to be essential to the development of science. By focusing attention upon a small range of relatively esoteric problems, the paradigm forces scientists to investigate some part of nature in a detail and depth that would otherwise be unimaginable. And normal science possesses a built-in mechanism that ensures the relaxation of the restrictions that bound research whenever the paradigm from which they derive ceases to function effectively. At that point scientists begin to behave differently, and the nature of their research problems changes. In the interim, however, during the

[1] Bernard Barber, "Resistance by Scientists to Scientific Discovery," *Science*, CXXXIV (1961), 596–602.

period when the paradigm is successful, the profession will have solved problems that its members could scarcely have imagined and would never have undertaken without commitment to the paradigm. And at least part of that achievement always proves to be permanent.

To display more clearly what is meant by normal or paradigm-based research, let me now attempt to classify and illustrate the problems of which normal science principally consists. For convenience I postpone theoretical activity and begin with fact-gathering, that is, with the experiments and observations described in the technical journals through which scientists inform their professional colleagues of the results of their continuing research. On what aspects of nature do scientists ordinarily report? What determines their choice? And, since most scientific observation consumes much time, equipment, and money, what motivates the scientist to pursue that choice to a conclusion?

There are, I think, only three normal foci for factual scientific investigation, and they are neither always nor permanently distinct. First is that class of facts that the paradigm has shown to be particularly revealing of the nature of things. By employing them in solving problems, the paradigm has made them worth determining both with more precision and in a larger variety of situations. At one time or another, these significant factual determinations have included: in astronomy—stellar position and magnitude, the periods of eclipsing binaries and of planets; in physics—the specific gravities and compressibilities of materials, wave lengths and spectral intensities, electrical conductivities and contact potentials; and in chemistry—composition and combining weights, boiling points and acidity of solutions, structural formulas and optical activities. Attempts to increase the accuracy and scope with which facts like these are known occupy a significant fraction of the literature of experimental and observational science. Again and again complex special apparatus has been designed for such purposes, and the invention, construction, and deployment of that apparatus have demanded first-rate talent, much time, and considerable financial backing. Synchrotrons and radiotelescopes are only the most recent examples of the lengths to which research workers will go if a paradigm assures them that the facts they seek are important. From Tycho Brahe to E. O. Lawrence, some scientists have acquired great reputations, not from any novelty of their discoveries, but from the precision, reliability, and scope of the methods they developed for the redetermination of a previously known sort of fact.

A second usual but smaller class of factual determinations is directed to those facts that, though often without much intrinsic interest, can be compared directly with predictions from the paradigm theory. As we shall see shortly, when I turn from the experimental to the theoretical problems of normal science, there are seldom many areas in which a scientific theory, particularly if it is cast in a predominantly mathematical form, can be directly compared with nature. No more than three such areas are even yet accessible to Einstein's general theory of relativity.[2] Furthermore, even in those areas where application is possible, it often demands theoretical and instrumental approximations that severely limit the agreement to be expected. Improving that agreement or finding new areas in which agreement can be demonstrated at all presents a constant challenge to the skill and imagination of the experimentalist and observer. Special telescopes to demonstrate the Copernican prediction of annual parallax; Atwood's machine, first invented almost a century after the *Principia,* to give the first unequivocal demonstration of Newton's second law; Foucault's apparatus to show that the speed of light is greater in air than in water; or the gigantic scintillation counter designed to demonstrate the existence of

[2] The only long-standing check point still generally recognized is the precession of Mercury's perihelion. The red shift in the spectrum of light from distant stars can be derived from considerations more elementary than general relativity, and the same may be possible for the bending of light around the sun, a point now in some dispute. In any case, measurements of the latter phenomenon remain equivocal. One additional check point may have been established very recently: the gravitational shift of Mossbauer radiation. Perhaps there will soon be others in this now active but long dormant field. For an up-to-date capsule account of the problem, see L. I. Schiff, "A Report on the NASA Conference on Experimental Tests of Theories of Relativity," *Physics Today,* XIV (1961), 42–48.

the neutrino—these pieces of special apparatus and many others like them illustrate the immense effort and ingenuity that have been required to bring nature and theory into closer and closer agreement.[3] That attempt to demonstrate agreement is a second type of normal experimental work, and it is even more obviously dependent than the first upon a paradigm. The existence of the paradigm sets the problem to be solved; often the paradigm theory is implicated directly in the design of apparatus able to solve the problem. Without the *Principia*, for example, measurements made with the Atwood machine would have meant nothing at all.

A third class of experiments and observations exhausts, I think, the fact-gathering activities of normal science. It consists of empirical work undertaken to articulate the paradigm theory, resolving some of its residual ambiguities and permitting the solution of problems to which it had previously only drawn attention. This class proves to be the most important of all, and its description demands its subdivision. In the more mathematical sciences, some of the experiments aimed at articulation are directed to the determination of physical constants. Newton's work, for example, indicated that the force between two unit masses at unit distance would be the same for all types of matter at all positions in the universe. But his own problems could be solved without even estimating the size of this attraction, the universal gravitational constant; and no one else devised apparatus able to determine it for a century after the *Principia* appeared. Nor was Cavendish's famous determination in the 1790's the last. Because of its central position in physical theory, improved values of the gravitational constant have been the object of repeated efforts ever since by a number of outstanding

[3] For two of the parallax telescopes, see Abraham Wolf, *A History of Science, Technology, and Philosophy in the Eighteenth Century* (2d ed.; London, 1952), pp. 103–5. For the Atwood machine, see N. R. Hanson, *Patterns of Discovery* (Cambridge, 1958), pp. 100–102, 207–8. For the last two pieces of special apparatus, see M. L. Foucault, "Méthode générale pour mesurer la vitesse de la lumière dans l'air et les milieux transparants. Vitesses relatives de la lumière dans l'air et dans l'eau . . . ," *Comptes rendus . . . de l'Académie des sciences,* XXX (1850), 551–60; and C. L. Cowan, Jr., *et al.*, "Detection of the Free Neutrino: A Confirmation," *Science,* CXXIV (1956), 103–4.

experimentalists.[4] Other examples of the same sort of continuing work would include determinations of the astronomical unit, Avogadro's number, Joule's coefficient, the electronic charge, and so on. Few of these elaborate efforts would have been conceived and none would have been carried out without a paradigm theory to define the problem and to guarantee the existence of a stable solution.

Efforts to articulate a paradigm are not, however, restricted to the determination of universal constants. They may, for example, also aim at quantitative laws: Boyle's Law relating gas pressure to volume, Coulomb's Law of electrical attraction, and Joule's formula relating heat generated to electrical resistance and current are all in this category. Perhaps it is not apparent that a paradigm is prerequisite to the discovery of laws like these. We often hear that they are found by examining measurements undertaken for their own sake and without theoretical commitment. But history offers no support for so excessively Baconian a method. Boyle's experiments were not conceivable (and if conceived would have received another interpretation or none at all) until air was recognized as an elastic fluid to which all the elaborate concepts of hydrostatics could be applied.[5] Coulomb's success depended upon his constructing special apparatus to measure the force between point charges. (Those who had previously measured electrical forces using ordinary pan balances, etc., had found no consistent or simple regularity at all.) But that design, in turn, depended upon the previous recognition that every particle of electric fluid acts upon every other at a distance. It was for the force between such particles—the only force which might safely be assumed

[4] J. H. P[oynting] reviews some two dozen measurements of the gravitational constant between 1741 and 1901 in "Gravitation Constant and Mean Density of the Earth," *Encyclopaedia Britannica* (11th ed.; Cambridge, 1910–11), XII, 385–89.

[5] For the full transplantation of hydrostatic concepts into pneumatics, see *The Physical Treatises of Pascal,* trans. I. H. B. Spiers and A. G. H. Spiers, with an introduction and notes by F. Barry (New York, 1937). Torricelli's original introduction of the parallelism ("We live submerged at the bottom of an ocean of the element air") occurs on p. 164. Its rapid development is displayed by the two main treatises.

a simple function of distance—that Coulomb was looking.[6] Joule's experiments could also be used to illustrate how quantitative laws emerge through paradigm articulation. In fact, so general and close is the relation between qualitative paradigm and quantitative law that, since Galileo, such laws have often been correctly guessed with the aid of a paradigm years before apparatus could be designed for their experimental determination.[7]

Finally, there is a third sort of experiment which aims to articulate a paradigm. More than the others this one can resemble exploration, and it is particularly prevalent in those periods and sciences that deal more with the qualitative than with the quantitative aspects of nature's regularity. Often a paradigm developed for one set of phenomena is ambiguous in its application to other closely related ones. Then experiments are necessary to choose among the alternative ways of applying the paradigm to the new area of interest. For example, the paradigm applications of the caloric theory were to heating and cooling by mixtures and by change of state. But heat could be released or absorbed in many other ways—e.g., by chemical combination, by friction, and by compression or absorption of a gas—and to each of these other phenomena the theory could be applied in several ways. If the vacuum had a heat capacity, for example, heating by compression could be explained as the result of mixing gas with void. Or it might be due to a change in the specific heat of gases with changing pressure. And there were several other explanations besides. Many experiments were undertaken to elaborate these various possibilities and to distinguish between them; all these experiments arose from the caloric theory as paradigm, and all exploited it in the design of experiments and in the interpretation of results.[8] Once the phe-

[6] Duane Roller and Duane H. D. Roller, *The Development of the Concept of Electric Charge: Electricity from the Greeks to Coulomb* ("Harvard Case Histories in Experimental Science," Case 8; Cambridge, Mass., 1954), pp. 66–80.

[7] For examples, see T. S. Kuhn, "The Function of Measurement in Modern Physical Science," *Isis,* LII (1961), 161–93.

[8] T. S. Kuhn, "The Caloric Theory of Adiabatic Compression," *Isis,* XLIX (1958), 132–40.

nomenon of heating by compression had been established, all further experiments in the area were paradigm-dependent in this way. Given the phenomenon, how else could an experiment to elucidate it have been chosen?

Turn now to the theoretical problems of normal science, which fall into very nearly the same classes as the experimental and observational. A part of normal theoretical work, though only a small part, consists simply in the use of existing theory to predict factual information of intrinsic value. The manufacture of astronomical ephemerides, the computation of lens characteristics, and the production of radio propagation curves are examples of problems of this sort. Scientists, however, generally regard them as hack work to be relegated to engineers or technicians. At no time do very many of them appear in significant scientific journals. But these journals do contain a great many theoretical discussions of problems that, to the non-scientist, must seem almost identical. These are the manipulations of theory undertaken, not because the predictions in which they result are intrinsically valuable, but because they can be confronted directly with experiment. Their purpose is to display a new application of the paradigm or to increase the precision of an application that has already been made.

The need for work of this sort arises from the immense difficulties often encountered in developing points of contact between a theory and nature. These difficulties can be briefly illustrated by an examination of the history of dynamics after Newton. By the early eighteenth century those scientists who found a paradigm in the *Principia* took the generality of its conclusions for granted, and they had every reason to do so. No other work known to the history of science has simultaneously permitted so large an increase in both the scope and precision of research. For the heavens Newton had derived Kepler's Laws of planetary motion and also explained certain of the observed respects in which the moon failed to obey them. For the earth he had derived the results of some scattered observations on pendulums and the tides. With the aid of additional but *ad hoc* assumptions, he had also been able to derive Boyle's Law and an important formula for the speed of sound in air. Given the state of science at the time, the success of the demonstrations was extremely impressive. Yet given the presumptive generality of Newton's Laws, the number of these applications was not great, and Newton developed almost no others. Furthermore, compared with what any graduate student of physics can achieve with those same laws today, Newton's few applications were not even developed with precision. Finally, the *Principia* had been designed for application chiefly to problems of celestial mechanics. How to adapt it for terrestrial applications, particularly for those of motion under constraint, was by no means clear. Terrestrial problems were, in any case, already being attacked with great success by a quite different set of techniques developed originally by Galileo and Huyghens and extended on the Continent during the eighteenth century by the Bernoullis, d'Alembert, and many others. Presumably their techniques and those of the *Principia* could be shown to be special cases of a more general formulation, but for some time no one saw quite how.[9]

Restrict attention for the moment to the problem of precision. We have already illustrated its empirical aspect. Special equipment—like Cavendish's apparatus, the Atwood machine, or improved telescopes—was required in order to provide the special data that the concrete applications of Newton's paradigm demanded. Similar difficulties in obtaining agreement existed on the side of theory. In applying his laws to pendulums, for example, Newton was forced to treat the bob as a mass point in order to provide a unique definition of pendulum length. Most of his theorems, the few exceptions being hypothetical and preliminary, also ignored the effect of air resistance. These were sound physical approximations. Nevertheless, as approximations they restricted the agreement to be expected

[9] C. Truesdell, "A Program toward Rediscovering the Rational Mechanics of the Age of Reason," *Archive for History of the Exact Sciences*, I (1960), 3–36, and "Reactions of Late Baroque Mechanics to Success, Conjecture, Error, and Failure in Newton's *Principia*," *Texas Quarterly*, X (1967), 281–97. T. L. Hankins, "The Reception of Newton's Second Law of Motion in the Eighteenth Century." *Archives internationales d'histoire des sciences*, XX (1967), 42–65.

between Newton's predictions and actual experiments. The same difficulties appear even more clearly in the application of Newton's theory to the heavens. Simple quantitative telescopic observations indicate that the planets do not quite obey Kepler's Laws, and Newton's theory indicates that they should not. To derive those laws, Newton had been forced to neglect all gravitational attraction except that between individual planets and the sun. Since the planets also attract each other, only approximate agreement between the applied theory and telescopic observation could be expected.[10]

The agreement obtained was, of course, more than satisfactory to those who obtained it. Excepting for some terrestrial problems, no other theory could do nearly so well. None of those who questioned the validity of Newton's work did so because of its limited agreement with experiment and observation. Nevertheless, these limitations of agreement left many fascinating theoretical problems for Newton's successors. Theoretical techniques were, for example, required for treating the motions of more than two simultaneously attracting bodies and for investigating the stability of perturbed orbits. Problems like these occupied many of Europe's best mathematicians during the eighteenth and early nineteenth century. Euler, Lagrange, Laplace, and Gauss all did some of their most brilliant work on problems aimed to improve the match between Newton's paradigm and observation of the heavens. Many of these figures worked simultaneously to develop the mathematics required for applications that neither Newton nor the contemporary Continental school of mechanics had even attempted. They produced, for example, an immense literature and some very powerful mathematical techniques for hydrodynamics and for the problem of vibrating strings. These problems of application account for what is probably the most brilliant and consuming scientific work of the eighteenth century. Other examples could be discovered by an examination of the post-paradigm period in the development of thermodynamics, the wave theory of light, electromagnetic the-

[10] Wolf, *op. cit.*, pp. 75–81, 96–101; and William Whewell, *History of the Inductive Sciences* (rev. ed.; London, 1847), II, 213–71.

ory, or any other branch of science whose fundamental laws are fully quantitative. At least in the more mathematical sciences, most theoretical work is of this sort.

But it is not all of this sort. Even in the mathematical sciences there are also theoretical problems of paradigm articulation; and during periods when scientific development is predominantly qualitative, these problems dominate. Some of the problems, in both the more quantitative and more qualitative sciences, aim simply at clarification by reformulation. The *Principia*, for example, did not always prove an easy work to apply, partly because it retained some of the clumsiness inevitable in a first venture and partly because so much of its meaning was only implicit in its applications. For many terrestrial applications, in any case, an apparently unrelated set of Continental techniques seemed vastly more powerful. Therefore, from Euler and Lagrange in the eighteenth century to Hamilton, Jacobi, and Hertz in the nineteenth, many of Europe's most brilliant mathematical physicists repeatedly endeavored to reformulate mechanical theory in an equivalent but logically and aesthetically more satisfying form. They wished, that is, to exhibit the explicit and implicit lessons of the *Principia* and of Continental mechanics in a logically more coherent version, one that would be at once more uniform and less equivocal in its application to the newly elaborated problems of mechanics.[11]

Similar reformulations of a paradigm have occurred repeatedly in all of the sciences, but most of them have produced more substantial changes in the paradigm than the reformulations of the *Principia* cited above. Such changes result from the empirical work previously described as aimed at paradigm articulation. Indeed, to classify that sort of work as empirical was arbitrary. More than any other sort of normal research, the problems of paradigm articulation are simultaneously theoretical and experimental; the examples given previously will serve equally well here. Before he could construct his equipment and make measurements with it, Coulomb had to employ electrical theory to determine how his equipment should be built. The

[11] René Dugas, *Histoire de la mécanique* (Neuchatel, 1950), Books IV–V.

consequence of his measurements was a refinement in that theory. Or again, the men who designed the experiments that were to distinguish between the various theories of heating by compression were generally the same men who had made up the versions being compared. They were working both with fact and with theory, and their work produced not simply new information but a more precise paradigm, obtained by the elimination of ambiguities that the original from which they worked had retained. In many sciences, most normal work is of this sort.

These three classes of problems—determination of significant fact, matching of facts with theory, and articulation of theory—exhaust, I think, the literature of normal science, both empirical and theoretical. They do not, of course, quite exhaust the entire literature of science. There are also extraordinary problems, and it may well be their resolution that makes the scientific enterprise as a whole so particularly worthwhile. But extraordinary problems are not to be had for the asking. They emerge only on special occasions prepared by the advance of normal research. Inevitably, therefore, the overwhelming majority of the problems undertaken by even the very best scientists usually fall into one of the three categories outlined above. Work under the paradigm can be conducted in no other way, and to desert the paradigm is to cease practicing the science it defines. We shall shortly discover that such desertions do occur. They are the pivots about which scientific revolutions turn. But before beginning the study of such revolutions, we require a more panoramic view of the normal-scientific pursuits that prepare the way.

IV. Normal Science as Puzzle-solving

Perhaps the most striking feature of the normal research problems we have just encountered is how little they aim to produce major novelties, conceptual or phenomenal. Sometimes, as in a wave-length measurement, everything but the most esoteric detail of the result is known in advance, and the typical latitude of expectation is only somewhat wider. Coulomb's measurements need not, perhaps, have fitted an inverse square law; the men who worked on heating by compression were often prepared for any one of several results. Yet even in cases like these the range of anticipated, and thus of assimilable, results is always small compared with the range that imagination can conceive. And the project whose outcome does not fall in that narrower range is usually just a research failure, one which reflects not on nature but on the scientist.

In the eighteenth century, for example, little attention was paid to the experiments that measured electrical attraction with devices like the pan balance. Because they yielded neither consistent nor simple results, they could not be used to articulate the paradigm from which they derived. Therefore, they remained *mere* facts, unrelated and unrelatable to the continuing progress of electrical research. Only in retrospect, possessed of a subsequent paradigm, can we see what characteristics of electrical phenomena they display. Coulomb and his contemporaries, of course, also possessed this later paradigm or one that, when applied to the problem of attraction, yielded the same expectations. That is why Coulomb was able to design apparatus that gave a result assimilable by paradigm articulation. But it is also why that result surprised no one and why several of Coulomb's contemporaries had been able to predict it in advance. Even the project whose goal is paradigm articulation does not aim at the *unexpected* novelty.

But if the aim of normal science is not major substantive novelties—if failure to come near the anticipated result is usually failure as a scientist—then why are these problems undertaken at all? Part of the answer has already been developed. To scientists, at least, the results gained in normal research are significant because they add to the scope and precision with which the paradigm can be applied. That answer, however, cannot account for the enthusiasm and devotion that scientists display for the problems of normal research. No one devotes years to, say, the development of a better spectrometer or the production of an improved solution to the problem of vibrating strings simply because of the importance of the information that will be obtained. The data to be gained by computing ephemerides or by further measurements with an existing instrument are often just as significant, but those activities are regularly spurned by scientists because they are so largely repetitions of procedures that have been carried through before. That rejection provides a clue to the fascination of the normal research problem. Though its outcome can be anticipated, often in detail so great that what remains to be known is itself uninteresting, the way to achieve that outcome remains very much in doubt. Bringing a normal research problem to a conclusion is achieving the anticipated in a new way, and it requires the solution of all sorts of complex instrumental, conceptual, and mathematical puzzles. The man who succeeds proves himself an expert puzzle-solver, and the challenge of the puzzle is an important part of what usually drives him on.

The terms 'puzzle' and 'puzzle-solver' highlight several of the themes that have become increasingly prominent in the preceding pages. Puzzles are, in the entirely standard meaning here employed, that special category of problems that can serve to test ingenuity or skill in solution. Dictionary illustrations are 'jigsaw puzzle' and 'crossword puzzle,' and it is the characteristics that these share with the problems of normal science that we now need to isolate. One of them has just been mentioned. It is no criterion of goodness in a puzzle that its outcome be intrinsically interesting or important. On the contrary, the really pressing problems, e.g., a cure for cancer or the design of a lasting peace, are often not puzzles at all, largely because they may not have any solution. Consider the jigsaw puzzle whose pieces are selected at random from each of two different puzzle boxes. Since that problem is likely to defy (though it might not) even the most ingenious of men, it cannot serve as a test of skill in solution. In any usual sense it is not a puzzle at all. Though intrinsic value is no criterion for a puzzle, the assured existence of a solution is.

We have already seen, however, that one of the things a scientific community acquires with a paradigm is a criterion for choosing problems that, while the paradigm is taken for granted, can be assumed to have solutions. To a great extent these are the only problems that the community will admit as scientific or encourage its members to undertake. Other problems, including many that had previously been standard, are rejected as metaphysical, as the concern of another discipline, or sometimes as just too problematic to be worth the time. A paradigm can, for that matter, even insulate the community from those socially important problems that are not reducible to the puzzle form, because they cannot be stated in terms of the conceptual and instrumental tools the paradigm supplies. Such problems can be a distraction, a lesson brilliantly illustrated by several facets of seventeenth-century Baconianism and by some of the contemporary social sciences. One of the reasons why normal science seems to progress so rapidly is that its practitioners concentrate on problems that only their own lack of ingenuity should keep them from solving.

If, however, the problems of normal science are puzzles in this sense, we need no longer ask why scientists attack them with such passion and devotion. A man may be attracted to science for all sorts of reasons. Among them are the desire to be useful, the excitement of exploring new territory, the hope of finding order, and the drive to test established knowledge. These motives and others besides also help to determine the particular problems that will later engage him. Furthermore, though the result is occasional frustration, there is good reason why motives like these should first attract him and then lead him on.[1] The scientific enterprise as a whole does from time to time prove useful, open up new territory, display order, and test long-accepted belief. Nevertheless, *the individual* engaged on a normal research problem *is almost never doing any one of these things.* Once engaged, his motivation is of a rather different sort. What then challenges him is the conviction that, if only he is skilful enough, he will succeed in solving a puzzle that no one before has solved or solved so well. Many of the greatest scientific minds have devoted all of their professional attention to demanding puzzles of this sort. On most occasions any particular field of specialization offers nothing else to do, a fact that makes it no less fascinating to the proper sort of addict.

Turn now to another, more difficult, and more revealing aspect of the parallelism between puzzles and the problems of normal science. If it is to classify as a puzzle, a problem must be characterized by more than an assured solution. There must also be rules that limit both the nature of acceptable solutions and the steps by which they are to be obtained. To solve a jigsaw puzzle is not, for example, merely "to make a picture." Either a child or a contemporary artist could do that by scattering selected pieces, as abstract shapes, upon some neutral ground. The picture thus produced might be far better, and would certainly be more original, than the one from which the puzzle had been made. Nevertheless, such a picture would not be a solution. To achieve that all the pieces must be used, their plain sides must be turned down, and they must be interlocked without forcing until no holes remain. Those are among the rules that govern jigsaw-puzzle solutions. Similar restrictions upon the admissible solutions of crossword puzzles, riddles, chess problems, and so on, are readily discovered.

If we can accept a considerably broadened use of the term

[1] The frustrations induced by the conflict between the individual's role and the over-all pattern of scientific development can, however, occasionally be quite serious. On this subject, see Lawrence S. Kubie, "Some Unsolved Problems of the Scientific Career," *American Scientist,* XLI (1953), 596–613; and XLII (1954), 104–12.

'rule'—one that will occasionally equate it with 'established viewpoint' or with 'preconception'—then the problems accessible within a given research tradition display something much like this set of puzzle characteristics. The man who builds an instrument to determine optical wave lengths must not be satisfied with a piece of equipment that merely attributes particular numbers to particular spectral lines. He is not just an explorer or measurer. On the contrary, he must show, by analyzing his apparatus in terms of the established body of optical theory, that the numbers his instrument produces are the ones that enter theory as wave lengths. If some residual vagueness in the theory or some unanalyzed component of his apparatus prevents his completing that demonstration, his colleagues may well conclude that he has measured nothing at all. For example, the electron-scattering maxima that were later diagnosed as indices of electron wave length had no apparent significance when first observed and recorded. Before they became measures of anything, they had to be related to a theory that predicted the wave-like behavior of matter in motion. And even after that relation was pointed out, the apparatus had to be redesigned so that the experimental results might be correlated unequivocally with theory.[2] Until those conditions had been satisfied, no problem had been solved.

Similar sorts of restrictions bound the admissible solutions to theoretical problems. Throughout the eighteenth century those scientists who tried to derive the observed motion of the moon from Newton's laws of motion and gravitation consistently failed to do so. As a result, some of them suggested replacing the inverse square law with a law that deviated from it at small distances. To do that, however, would have been to change the paradigm, to define a new puzzle, and not to solve the old one. In the event, scientists preserved the rules until, in 1750, one of them discovered how they could successfully be applied.[3]

[2] For a brief account of the evolution of these experiments, see page 4 of C. J. Davisson's lecture in *Les prix Nobel en 1937* (Stockholm, 1938).

[3] W. Whewell, *History of the Inductive Sciences* (rev. ed.; London, 1847), II, 101–5, 220–22.

Only a change in the rules of the game could have provided an alternative.

The study of normal-scientific traditions discloses many additional rules, and these provide much information about the commitments that scientists derive from their paradigms. What can we say are the main categories into which these rules fall?[4] The most obvious and probably the most binding is exemplified by the sorts of generalizations we have just noted. These are explicit statements of scientific law and about scientific concepts and theories. While they continue to be honored, such statements help to set puzzles and to limit acceptable solutions. Newton's Laws, for example, performed those functions during the eighteenth and nineteenth centuries. As long as they did so, quantity-of-matter was a fundamental ontological category for physical scientists, and the forces that act between bits of matter were a dominant topic for research.[5] In chemistry the laws of fixed and definite proportions had, for a long time, an exactly similar force—setting the problem of atomic weights, bounding the admissible results of chemical analyses, and informing chemists what atoms and molecules, compounds and mixtures were.[6] Maxwell's equations and the laws of statistical thermodynamics have the same hold and function today.

Rules like these are, however, neither the only nor even the most interesting variety displayed by historical study. At a level lower or more concrete than that of laws and theories, there is, for example, a multitude of commitments to preferred types of instrumentation and to the ways in which accepted instruments may legitimately be employed. Changing attitudes toward the role of fire in chemical analyses played a vital part in the de-

[4] I owe this question to W. O. Hagstrom, whose work in the sociology of science sometimes overlaps my own.

[5] For these aspects of Newtonianism, see I. B. Cohen, *Franklin and Newton: An Inquiry into Speculative Newtonian Experimental Science and Franklin's Work in Electricity as an Example Thereof* (Philadelphia, 1956), chap. vii, esp. pp. 255–57, 275–77.

[6] This example is discussed at length near the end of Section X.

velopment of chemistry in the seventeenth century.[7] Helmholtz, in the nineteenth, encountered strong resistance from physiologists to the notion that physical experimentation could illuminate their field.[8] And in this century the curious history of chemical chromatography again illustrates the endurance of instrumental commitments that, as much as laws and theory, provide scientists with rules of the game.[9] When we analyze the discovery of X-rays, we shall find reasons for commitments of this sort.

Less local and temporary, though still not unchanging characteristics of science, are the higher level, quasi-metaphysical commitments that historical study so regularly displays. After about 1630, for example, and particularly after the appearance of Descartes's immensely influential scientific writings, most physical scientists assumed that the universe was composed of microscopic corpuscles and that all natural phenomena could be explained in terms of corpuscular shape, size, motion, and interaction. That nest of commitments proved to be both metaphysical and methodological. As metaphysical, it told scientists what sorts of entities the universe did and did not contain: there was only shaped matter in motion. As methodological, it told them what ultimate laws and fundamental explanations must be like: laws must specify corpuscular motion and interaction, and explanation must reduce any given natural phenomenon to corpuscular action under these laws. More important still, the corpuscular conception of the universe told scientists what many of their research problems should be. For example, a chemist who, like Boyle, embraced the new philosophy gave particular attention to reactions that could be viewed as transmutations. More clearly than any others these displayed the process of corpuscular rearrangement that must underlie all

[7] H. Metzger, *Les doctrines chimiques en France du début du XVIIe siècle à la fin du XVIIIe siècle* (Paris, 1923), pp. 359–61; Marie Boas, *Robert Boyle and Seventeenth-Century Chemistry* (Cambridge, 1958), pp. 112–15.

[8] Leo Königsberger, *Hermann von Helmholtz*, trans. Francis A. Welby (Oxford, 1906), pp. 65–66.

[9] James E. Meinhard, "Chromatography: A Perspective," *Science*, CX (1949), 387–92.

chemical change.[10] Similar effects of corpuscularism can be observed in the study of mechanics, optics, and heat.

Finally, at a still higher level, there is another set of commitments without which no man is a scientist. The scientist must, for example, be concerned to understand the world and to extend the precision and scope with which it has been ordered. That commitment must, in turn, lead him to scrutinize, either for himself or through colleagues, some aspect of nature in great empirical detail. And, if that scrutiny displays pockets of apparent disorder, then these must challenge him to a new refinement of his observational techniques or to a further articulation of his theories. Undoubtedly there are still other rules like these, ones which have held for scientists at all times.

The existence of this strong network of commitments—conceptual, theoretical, instrumental, and methodological—is a principal source of the metaphor that relates normal science to puzzle-solving. Because it provides rules that tell the practitioner of a mature specialty what both the world and his science are like, he can concentrate with assurance upon the esoteric problems that these rules and existing knowledge define for him. What then personally challenges him is how to bring the residual puzzle to a solution. In these and other respects a discussion of puzzles and of rules illuminates the nature of normal scientific practice. Yet, in another way, that illumination may be significantly misleading. Though there obviously are rules to which all the practitioners of a scientific specialty adhere at a given time, those rules may not by themselves specify all that the practice of those specialists has in common. Normal science is a highly determined activity, but it need not be entirely determined by rules. That is why, at the start of this essay, I introduced shared paradigms rather than shared rules, assumptions, and points of view as the source of coherence for normal research traditions. Rules, I suggest, derive from paradigms, but paradigms can guide research even in the absence of rules.

[10] For corpuscularism in general, see Marie Boas, "The Establishment of the Mechanical Philosophy," *Osiris*, X (1952), 412–541. For its effects on Boyle's chemistry, see T. S. Kuhn, "Robert Boyle and Structural Chemistry in the Seventeenth Century," *Isis*, XLIII (1952), 12–36.

V. The Priority of Paradigms

To discover the relation between rules, paradigms, and normal science, consider first how the historian isolates the particular loci of commitment that have just been described as accepted rules. Close historical investigation of a given specialty at a given time discloses a set of recurrent and quasi-standard illustrations of various theories in their conceptual, observational, and instrumental applications. These are the community's paradigms, revealed in its textbooks, lectures, and laboratory exercises. By studying them and by practicing with them, the members of the corresponding community learn their trade. The historian, of course, will discover in addition a penumbral area occupied by achievements whose status is still in doubt, but the core of solved problems and techniques will usually be clear. Despite occasional ambiguities, the paradigms of a mature scientific community can be determined with relative ease.

The determination of shared paradigms is not, however, the determination of shared rules. That demands a second step and one of a somewhat different kind. When undertaking it, the historian must compare the community's paradigms with each other and with its current research reports. In doing so, his object is to discover what isolable elements, explicit or implicit, the members of that community may have *abstracted* from their more global paradigms and deployed as rules in their research. Anyone who has attempted to describe or analyze the evolution of a particular scientific tradition will necessarily have sought accepted principles and rules of this sort. Almost certainly, as the preceding section indicates, he will have met with at least partial success. But, if his experience has been at all like my own, he will have found the search for rules both more difficult and less satisfying than the search for paradigms. Some of the generalizations he employs to describe the community's shared beliefs will present no problems. Others, however, including some of those used as illustrations above, will seem a shade too strong. Phrased in just that way, or in any other way he can imagine, they would almost certainly have been rejected by some members of the group he studies. Nevertheless, if the coherence of the research tradition is to be understood in terms of rules, some specification of common ground in the corresponding area is needed. As a result, the search for a body of rules competent to constitute a given normal research tradition becomes a source of continual and deep frustration.

Recognizing that frustration, however, makes it possible to diagnose its source. Scientists can agree that a Newton, Lavoisier, Maxwell, or Einstein has produced an apparently permanent solution to a group of outstanding problems and still disagree, sometimes without being aware of it, about the particular abstract characteristics that make those solutions permanent. They can, that is, agree in their *identification* of a paradigm without agreeing on, or even attempting to produce, a full *interpretation* or *rationalization* of it. Lack of a standard interpretation or of an agreed reduction to rules will not prevent a paradigm from guiding research. Normal science can be determined in part by the direct inspection of paradigms, a process that is often aided by but does not depend upon the formulation of rules and assumptions. Indeed, the existence of a paradigm need not even imply that any full set of rules exists.[1]

Inevitably, the first effect of those statements is to raise problems. In the absence of a competent body of rules, what restricts the scientist to a particular normal-scientific tradition? What can the phrase 'direct inspection of paradigms' mean? Partial answers to questions like these were developed by the the late Ludwig Wittgenstein, though in a very different context. Because that context is both more elementary and more familiar, it will help to consider his form of the argument first. What need we know, Wittgenstein asked, in order that we

[1] Michael Polanyi has brilliantly developed a very similar theme, arguing that much of the scientist's success depends upon "tacit knowledge," i.e., upon knowledge that is acquired through practice and that cannot be articulated explicitly. See his *Personal Knowledge* (Chicago, 1958), particularly chaps. v and vi.

apply terms like 'chair,' or 'leaf,' or 'game' unequivocally and without provoking argument?[2]

That question is very old and has generally been answered by saying that we must know, consciously or intuitively, what a chair, or leaf, or game *is*. We must, that is, grasp some set of attributes that all games and that only games have in common. Wittgenstein, however, concluded that, given the way we use language and the sort of world to which we apply it, there need be no such set of characteristics. Though a discussion of *some* of the attributes shared by a *number* of games or chairs or leaves often helps us learn how to employ the corresponding term, there is no set of characteristics that is simultaneously applicable to all members of the class and to them alone. Instead, confronted with a previously unobserved activity, we apply the term 'game' because what we are seeing bears a close "family resemblance" to a number of the activities that we have previously learned to call by that name. For Wittgenstein, in short, games, and chairs, and leaves are natural families, each constituted by a network of overlapping and crisscross resemblances. The existence of such a network sufficiently accounts for our success in identifying the corresponding object or activity. Only if the families we named overlapped and merged gradually into one another—only, that is, if there were no *natural* families—would our success in identifying and naming provide evidence for a set of common characteristics corresponding to each of the class names we employ.

Something of the same sort may very well hold for the various research problems and techniques that arise within a single normal-scientific tradition. What these have in common is not that they satisfy some explicit or even some fully discoverable set of rules and assumptions that gives the tradition its character and its hold upon the scientific mind. Instead, they may relate by resemblance and by modeling to one or another part of the scientific corpus which the community in question al-

[2] Ludwig Wittgenstein, *Philosophical Investigations*, trans. G. E. M. Anscombe (New York, 1953), pp. 31–36. Wittgenstein, however, says almost nothing about the sort of world necessary to support the naming procedure he outlines. Part of the point that follows cannot therefore be attributed to him.

ready recognizes as among its established achievements. Scientists work from models acquired through education and through subsequent exposure to the literature often without quite knowing or needing to know what characteristics have given these models the status of community paradigms. And because they do so, they need no full set of rules. The coherence displayed by the research tradition in which they participate may not imply even the existence of an underlying body of rules and assumptions that additional historical or philosophical investigation might uncover. That scientists do not usually ask or debate what makes a particular problem or solution legitimate tempts us to suppose that, at least intuitively, they know the answer. But it may only indicate that neither the question nor the answer is felt to be relevant to their research. Paradigms may be prior to, more binding, and more complete than any set of rules for research that could be unequivocally abstracted from them.

So far this point has been entirely theoretical: paradigms *could* determine normal science without the intervention of discoverable rules. Let me now try to increase both its clarity and urgency by indicating some of the reasons for believing that paradigms actually do operate in this manner. The first, which has already been discussed quite fully, is the severe difficulty of discovering the rules that have guided particular normal-scientific traditions. That difficulty is very nearly the same as the one the philosopher encounters when he tries to say what all games have in common. The second, to which the first is really a corollary, is rooted in the nature of scientific education. Scientists, it should already be clear, never learn concepts, laws, and theories in the abstract and by themselves. Instead, these intellectual tools are from the start encountered in a historically and pedagogically prior unit that displays them with and through their applications. A new theory is always announced together with applications to some concrete range of natural phenomena; without them it would not be even a candidate for acceptance. After it has been accepted, those same applications or others accompany the theory into the textbooks from which the future practitioner will learn his trade. They are not there merely as embroidery or even as documentation. On the contrary, the process of learning a theory depends upon the study of applications, including practice problem-solving both with a pencil and paper and with instruments in the laboratory. If, for example, the student of Newtonian dynamics ever discovers the meaning of terms like 'force,' 'mass,' 'space,' and 'time,' he does so less from the incomplete though sometimes helpful definitions in his text than by observing and participating in the application of these concepts to problem-solution.

That process of learning by finger exercise or by doing continues throughout the process of professional initiation. As the student proceeds from his freshman course to and through his doctoral dissertation, the problems assigned to him become more complex and less completely precedented. But they continue to be closely modeled on previous achievements as are the problems that normally occupy him during his subsequent independent scientific career. One is at liberty to suppose that somewhere along the way the scientist has intuitively abstracted rules of the game for himself, but there is little reason to believe it. Though many scientists talk easily and well about the particular individual hypotheses that underlie a concrete piece of current research, they are little better than laymen at characterizing the established bases of their field, its legitimate problems and methods. If they have learned such abstractions at all, they show it mainly through their ability to do successful research. That ability can, however, be understood without recourse to hypothetical rules of the game.

These consequences of scientific education have a converse that provides a third reason to suppose that paradigms guide research by direct modeling as well as through abstracted rules. Normal science can proceed without rules only so long as the relevant scientific community accepts without question the particular problem-solutions already achieved. Rules should therefore become important and the characteristic unconcern about them should vanish whenever paradigms or models are felt to be insecure. That is, moreover, exactly what does occur. The preparadigm period, in particular, is regularly marked by frequent and deep debates over legitimate methods, problems, and standards of solution, though these serve rather to define schools than to produce agreement. We have already noted a few of these debates in optics and electricity, and they played an even larger role in the development of seventeenth-century chemistry and of early nineteenth-century geology.[3] Furthermore, debates like these do not vanish once and for all with the appearance of a paradigm. Though almost non-existent during periods of normal science, they recur regularly just before and during scientific revolutions, the periods when paradigms are first under attack and then subject to change. The transition from Newtonian to quantum mechanics evoked many debates about both the nature and the standards of physics, some of which still continue.[4] There are people alive today who can remember the similar arguments engendered by Maxwell's electromagnetic theory and by statistical mechanics.[5] And earlier still, the assimilation of Galileo's and Newton's mechanics gave rise to a particularly famous series of debates with Aristotelians, Cartesians, and Leibnizians about the standards legitimate to science.[6] When scientists disagree about whether the fundamental problems of their field have been solved, the search for rules gains a function that it does not ordinarily possess. While

[3] For chemistry, see H. Metzger, *Les doctrines chimiques en France du début du XVIIe à la fin du XVIIIe siècle* (Paris, 1923), pp. 24–27, 146–49; and Marie Boas, *Robert Boyle and Seventeenth-Century Chemistry* (Cambridge, 1958), chap. ii. For geology, see Walter F. Cannon, "The Uniformitarian-Catastrophist Debate," *Isis*, LI (1960), 38–55; and C. C. Gillispie, *Genesis and Geology* (Cambridge, Mass., 1951), chaps. iv–v.

[4] For controversies over quantum mechanics, see Jean Ullmo, *La crise de la physique quantique* (Paris, 1950), chap. ii.

[5] For statistical mechanics, see René Dugas, *La théorie physique au sens de Boltzmann et ses prolongements modernes* (Neuchatel, 1959), pp. 158–84, 206–19. For the reception of Maxwell's work, see Max Planck, "Maxwell's Influence in Germany," in *James Clerk Maxwell: A Commemoration Volume, 1831–1931* (Cambridge, 1931), pp. 45–65, esp. pp. 58–63; and Silvanus P. Thompson, *The Life of William Thomson Baron Kelvin of Largs* (London, 1910), II, 1021–27.

[6] For a sample of the battle with the Aristotelians, see A. Koyré, "A Documentary History of the Problem of Fall from Kepler to Newton," *Transactions of the American Philosophical Society*, XLV (1955), 329–95. For the debates with the Cartesians and Leibnizians, see Pierre Brunet, *L'introduction des théories de Newton en France au XVIIIe siècle* (Paris, 1931); and A. Koyré, *From the Closed World to the Infinite Universe* (Baltimore, 1957), chap. xi.

paradigms remain secure, however, they can function without agreement over rationalization or without any attempted rationalization at all.

A fourth reason for granting paradigms a status prior to that of shared rules and assumptions can conclude this section. The introduction to this essay suggested that there can be small revolutions as well as large ones, that some revolutions affect only the members of a professional subspecialty, and that for such groups even the discovery of a new and unexpected phenomenon may be revolutionary. The next section will introduce selected revolutions of that sort, and it is still far from clear how they can exist. If normal science is so rigid and if scientific communities are so close-knit as the preceding discussion has implied, how can a change of paradigm ever affect only a small subgroup? What has been said so far may have seemed to imply that normal science is a single monolithic and unified enterprise that must stand or fall with any one of its paradigms as well as with all of them together. But science is obviously seldom or never like that. Often, viewing all fields together, it seems instead a rather ramshackle structure with little coherence among its various parts. Nothing said to this point should, however, conflict with that very familiar observation. On the contrary, substituting paradigms for rules should make the diversity of scientific fields and specialties easier to understand. Explicit rules, when they exist, are usually common to a very broad scientific group, but paradigms need not be. The practitioners of widely separated fields, say astronomy and taxonomic botany, are educated by exposure to quite different achievements described in very different books. And even men who, being in the same or in closely related fields, begin by studying many of the same books and achievements may acquire rather different paradigms in the course of professional specialization.

Consider, for a single example, the quite large and diverse community constituted by all physical scientists. Each member of that group today is taught the laws of, say, quantum mechanics, and most of them employ these laws at some point in their research or teaching. But they do not all learn the same applications of these laws, and they are not therefore all affected in the same ways by changes in quantum-mechanical practice. On the road to professional specialization, a few physical scientists encounter only the basic principles of quantum mechanics. Others study in detail the paradigm applications of these principles to chemistry, still others to the physics of the solid state, and so on. What quantum mechanics means to each of them depends upon what courses he has had, what texts he has read, and which journals he studies. It follows that, though a change in quantum-mechanical law will be revolutionary for all of these groups, a change that reflects only on one or another of the paradigm applications of quantum mechanics need be revolutionary only for the members of a particular professional subspecialty. For the rest of the profession and for those who practice other physical sciences, that change need not be revolutionary at all. In short, though quantum mechanics (or Newtonian dynamics, or electromagnetic theory) is a paradigm for many scientific groups, it is not the same paradigm for them all. Therefore, it can simultaneously determine several traditions of normal science that overlap without being coextensive. A revolution produced within one of these traditions will not necessarily extend to the others as well.

One brief illustration of specialization's effect may give this whole series of points additional force. An investigator who hoped to learn something about what scientists took the atomic theory to be asked a distinguished physicist and an eminent chemist whether a single atom of helium was or was not a molecule. Both answered without hesitation, but their answers were not the same. For the chemist the atom of helium was a molecule because it behaved like one with respect to the kinetic theory of gases. For the physicist, on the other hand, the helium atom was not a molecule because it displayed no molecular spectrum.[7] Presumably both men were talking of the same par-

[7] The investigator was James K. Senior, to whom I am indebted for a verbal report. Some related issues are treated in his paper, "The Vernacular of the Laboratory," *Philosophy of Science*, XXV (1958), 163–68.

ticle, but they were viewing it through their own research training and practice. Their experience in problem-solving told them what a molecule must be. Undoubtedly their experiences had had much in common, but they did not, in this case, tell the two specialists the same thing. As we proceed we shall discover how consequential paradigm differences of this sort can occasionally be.

VI. Anomaly and the Emergence of Scientific Discoveries

Normal science, the puzzle-solving activity we have just examined, is a highly cumulative enterprise, eminently successful in its aim, the steady extension of the scope and precision of scientific knowledge. In all these respects it fits with great precision the most usual image of scientific work. Yet one standard product of the scientific enterprise is missing. Normal science does not aim at novelties of fact or theory and, when successful, finds none. New and unsuspected phenomena are, however, repeatedly uncovered by scientific research, and radical new theories have again and again been invented by scientists. History even suggests that the scientific enterprise has developed a uniquely powerful technique for producing surprises of this sort. If this characteristic of science is to be reconciled with what has already been said, then research under a paradigm must be a particularly effective way of inducing paradigm change. That is what fundamental novelties of fact and theory do. Produced inadvertently by a game played under one set of rules, their assimilation requires the elaboration of another set. After they have become parts of science, the enterprise, at least of those specialists in whose particular field the novelties lie, is never quite the same again.

We must now ask how changes of this sort can come about, considering first discoveries, or novelties of fact, and then inventions, or novelties of theory. That distinction between discovery and invention or between fact and theory will, however, immediately prove to be exceedingly artificial. Its artificiality is an important clue to several of this essay's main theses. Examining selected discoveries in the rest of this section, we shall quickly find that they are not isolated events but extended episodes with a regularly recurrent structure. Discovery commences with the awareness of anomaly, i.e., with the recognition that nature has somehow violated the paradigm-induced expectations that govern normal science. It then continues with a more or less extended exploration of the area of anomaly. And it closes only when the paradigm theory has been adjusted so that the anomalous has become the expected. Assimilating a new sort of fact demands a more than additive adjustment of theory, and until that adjustment is completed—until the scientist has learned to see nature in a different way—the new fact is not quite a scientific fact at all.

To see how closely factual and theoretical novelty are intertwined in scientific discovery examine a particularly famous example, the discovery of oxygen. At least three different men have a legitimate claim to it, and several other chemists must, in the early 1770's, have had enriched air in a laboratory vessel without knowing it.[1] The progress of normal science, in this case of pneumatic chemistry, prepared the way to a breakthrough quite thoroughly. The earliest of the claimants to prepare a relatively pure sample of the gas was the Swedish apothecary, C. W. Scheele. We may, however, ignore his work since it was not published until oxygen's discovery had repeatedly been announced elsewhere and thus had no effect upon the historical pattern that most concerns us here.[2] The second in time to establish a claim was the British scientist and divine, Joseph Priestley, who collected the gas released by heated red oxide of mercury as one item in a prolonged normal investigation of the "airs" evolved by a large number of solid substances. In 1774 he identified the gas thus produced as nitrous oxide and in 1775, led by further tests, as common air with less than its usual quantity of phlogiston. The third claimant, Lavoisier, started the work that led him to oxygen after Priestley's experiments of 1774 and possibly as the result of a hint from Priestley. Early in

[1] For the still classic discussion of oxygen's discovery, see A. N. Meldrum, *The Eighteenth-Century Revolution in Science—the First Phase* (Calcutta, 1930), chap. v. An indispensable recent review, including an account of the priority controversy, is Maurice Daumas, *Lavoisier, théoricien et expérimentateur* (Paris, 1955), chaps. ii–iii. For a fuller account and bibliography, see also T. S. Kuhn, "The Historical Structure of Scientific Discovery," *Science*, CXXXVI (June 1, 1962), 760–64.

[2] See, however, Uno Bocklund, "A Lost Letter from Scheele to Lavoisier," *Lychnos*, 1957–58, pp. 39–62, for a different evaluation of Scheele's role.

1775 Lavoisier reported that the gas obtained by heating the red oxide of mercury was "air itself entire without alteration [except that] . . . it comes out more pure, more respirable."[3] By 1777, probably with the assistance of a second hint from Priestley, Lavoisier had concluded that the gas was a distinct species, one of the two main constituents of the atmosphere, a conclusion that Priestley was never able to accept.

This pattern of discovery raises a question that can be asked about every novel phenomenon that has ever entered the consciousness of scientists. Was it Priestley or Lavoisier, if either, who first discovered oxygen? In any case, when was oxygen discovered? In that form the question could be asked even if only one claimant had existed. As a ruling about priority and date, an answer does not at all concern us. Nevertheless, an attempt to produce one will illuminate the nature of discovery, because there is no answer of the kind that is sought. Discovery is not the sort of process about which the question is appropriately asked. The fact that it is asked—the priority for oxygen has repeatedly been contested since the 1780's—is a symptom of something askew in the image of science that gives discovery so fundamental a role. Look once more at our example. Priestley's claim to the discovery of oxygen is based upon his priority in isolating a gas that was later recognized as a distinct species. But Priestley's sample was not pure, and, if holding impure oxygen in one's hands is to discover it, that had been done by everyone who ever bottled atmospheric air. Besides, if Priestley was the discoverer, when was the discovery made? In 1774 he thought he had obtained nitrous oxide, a species he already knew; in 1775 he saw the gas as dephlogisticated air, which is still not oxygen or even, for phlogistic chemists, a quite unexpected sort of gas. Lavoisier's claim may be stronger, but it presents the same problems. If we refuse the palm to Priestley, we cannot award it to Lavoisier for the work of 1775 which led

[3] J. B. Conant, *The Overthrow of the Phlogiston Theory: The Chemical Revolution of 1775–1789* ("Harvard Case Histories in Experimental Science," Case 2; Cambridge, Mass., 1950), p. 23. This very useful pamphlet reprints many of the relevant documents.

him to identify the gas as the "air itself entire." Presumably we wait for the work of 1776 and 1777 which led Lavoisier to see not merely the gas but what the gas was. Yet even this award could be questioned, for in 1777 and to the end of his life Lavoisier insisted that oxygen was an atomic "principle of acidity" and that oxygen gas was formed only when that "principle" united with caloric, the matter of heat.[4] Shall we therefore say that oxygen had not yet been discovered in 1777? Some may be tempted to do so. But the principle of acidity was not banished from chemistry until after 1810, and caloric lingered until the 1860's. Oxygen had become a standard chemical substance before either of those dates.

Clearly we need a new vocabulary and concepts for analyzing events like the discovery of oxygen. Though undoubtedly correct, the sentence, "Oxygen was discovered," misleads by suggesting that discovering something is a single simple act assimilable to our usual (and also questionable) concept of seeing. That is why we so readily assume that discovering, like seeing or touching, should be unequivocally attributable to an individual and to a moment in time. But the latter attribution is always impossible, and the former often is as well. Ignoring Scheele, we can safely say that oxygen had not been discovered before 1774, and we would probably also say that it had been discovered by 1777 or shortly thereafter. But within those limits or others like them, any attempt to date the discovery must inevitably be arbitrary because discovering a new sort of phenomenon is necessarily a complex event, one which involves recognizing both *that* something is and *what* it is. Note, for example, that if oxygen were dephlogisticated air for us, we should insist without hesitation that Priestley had discovered it, though we would still not know quite when. But if both observation and conceptualization, fact and assimilation to theory, are inseparably linked in discovery, then discovery is a process and must take time. Only when all the relevant conceptual categories are prepared in advance, in which case the phenomenon would not

[4] H. Metzger, *La philosophie de la matière chez Lavoisier* (Paris, 1935); and Daumas, *op. cit.*, chap. vii.

be of a new sort, can discovering *that* and discovering *what* occur effortlessly, together, and in an instant.

Grant now that discovery involves an extended, though not necessarily long, process of conceptual assimilation. Can we also say that it involves a change in paradigm? To that question, no general answer can yet be given, but in this case at least, the answer must be yes. What Lavoisier announced in his papers from 1777 on was not so much the discovery of oxygen as the oxygen theory of combustion. That theory was the keystone for a reformulation of chemistry so vast that it is usually called the chemical revolution. Indeed, if the discovery of oxygen had not been an intimate part of the emergence of a new paradigm for chemistry, the question of priority from which we began would never have seemed so important. In this case as in others, the value placed upon a new phenomenon and thus upon its discoverer varies with our estimate of the extent to which the phenomenon violated paradigm-induced anticipations. Notice, however, since it will be important later, that the discovery of oxygen was not by itself the cause of the change in chemical theory. Long before he played any part in the discovery of the new gas, Lavoisier was convinced both that something was wrong with the phlogiston theory and that burning bodies absorbed some part of the atmosphere. That much he had recorded in a sealed note deposited with the Secretary of the French Academy in 1772.[5] What the work on oxygen did was to give much additional form and structure to Lavoisier's earlier sense that something was amiss. It told him a thing he was already prepared to discover—the nature of the substance that combustion removes from the atmosphere. That advance awareness of difficulties must be a significant part of what enabled Lavoisier to see in experiments like Priestley's a gas that Priestley had been unable to see there himself. Conversely, the fact that a major paradigm revision was needed to see what Lavoisier saw must be the principal reason why Priestley was, to the end of his long life, unable to see it.

[5] The most authoritative account of the origin of Lavoisier's discontent is Henry Guerlac, *Lavoisier—the Crucial Year: The Background and Origin of His First Experiments on Combustion in 1772* (Ithaca, N.Y., 1961).

Two other and far briefer examples will reinforce much that has just been said and simultaneously carry us from an elucidation of the nature of discoveries toward an understanding of the circumstances under which they emerge in science. In an effort to represent the main ways in which discoveries can come about, these examples are chosen to be different both from each other and from the discovery of oxygen. The first, X-rays, is a classic case of discovery through accident, a type that occurs more frequently than the impersonal standards of scientific reporting allow us easily to realize. Its story opens on the day that the physicist Roentgen interrupted a normal investigation of cathode rays because he had noticed that a barium platinocyanide screen at some distance from his shielded apparatus glowed when the discharge was in process. Further investigations—they required seven hectic weeks during which Roentgen rarely left the laboratory—indicated that the cause of the glow came in straight lines from the cathode ray tube, that the radiation cast shadows, could not be deflected by a magnet, and much else besides. Before announcing his discovery, Roentgen had convinced himself that his effect was not due to cathode rays but to an agent with at least some similarity to light.[6]

Even so brief an epitome reveals striking resemblances to the discovery of oxygen: before experimenting with red oxide of mercury, Lavoisier had performed experiments that did not produce the results anticipated under the phlogiston paradigm; Roentgen's discovery commenced with the recognition that his screen glowed when it should not. In both cases the perception of anomaly—of a phenomenon, that is, for which his paradigm had not readied the investigator—played an essential role in preparing the way for perception of novelty. But, again in both cases, the perception that something had gone wrong was only the prelude to discovery. Neither oxygen nor X-rays emerged without a further process of experimentation and assimilation. At what point in Roentgen's investigation, for example, ought we say that X-rays had actually been discovered? Not, in any

[6] L. W. Taylor, *Physics, the Pioneer Science* (Boston, 1941), pp. 790–94; and T. W. Chalmers, *Historic Researches* (London, 1949), pp. 218–19.

case, at the first instant, when all that had been noted was a glowing screen. At least one other investigator had seen that glow and, to his subsequent chagrin, discovered nothing at all.[7] Nor, it is almost as clear, can the moment of discovery be pushed forward to a point during the last week of investigation, by which time Roentgen was exploring the properties of the new radiation he had *already* discovered. We can only say that X-rays emerged in Würzburg between November 8 and December 28, 1895.

In a third area, however, the existence of significant parallels between the discoveries of oxygen and of X-rays is far less apparent. Unlike the discovery of oxygen, that of X-rays was not, at least for a decade after the event, implicated in any obvious upheaval in scientific theory. In what sense, then, can the assimilation of that discovery be said to have necessitated paradigm change? The case for denying such a change is very strong. To be sure, the paradigms subscribed to by Roentgen and his contemporaries could not have been used to predict X-rays. (Maxwell's electromagnetic theory had not yet been accepted everywhere, and the particulate theory of cathode rays was only one of several current speculations.) But neither did those paradigms, at least in any obvious sense, prohibit the existence of X-rays as the phlogiston theory had prohibited Lavoisier's interpretation of Priestley's gas. On the contrary, in 1895 accepted scientific theory and practice admitted a number of forms of radiation—visible, infrared, and ultraviolet. Why could not X-rays have been accepted as just one more form of a well-known class of natural phenomena? Why were they not, for example, received in the same way as the discovery of an additional chemical element? New elements to fill empty places in the periodic table were still being sought and found in Roentgen's day. Their pursuit was a standard project for normal science, and success was an occasion only for congratulations, not for surprise.

[7] E. T. Whittaker, *A History of the Theories of Aether and Electricity*, I (2d ed.; London, 1951), 358, n. 1. Sir George Thomson has informed me of a second near miss. Alerted by unaccountably fogged photographic plates, Sir William Crookes was also on the track of the discovery.

X-rays, however, were greeted not only with surprise but with shock. Lord Kelvin at first pronounced them an elaborate hoax.[8] Others, though they could not doubt the evidence, were clearly staggered by it. Though X-rays were not prohibited by established theory, they violated deeply entrenched expectations. Those expectations, I suggest, were implicit in the design and interpretation of established laboratory procedures. By the 1890's cathode ray equipment was widely deployed in numerous European laboratories. If Roentgen's apparatus had produced X-rays, then a number of other experimentalists must for some time have been producing those rays without knowing it. Perhaps those rays, which might well have other unacknowledged sources too, were implicated in behavior previously explained without reference to them. At the very least, several sorts of long familiar apparatus would in the future have to be shielded with lead. Previously completed work on normal projects would now have to be done again because earlier scientists had failed to recognize and control a relevant variable. X-rays, to be sure, opened up a new field and thus added to the potential domain of normal science. But they also, and this is now the more important point, changed fields that had already existed. In the process they denied previously paradigmatic types of instrumentation their right to that title.

In short, consciously or not, the decision to employ a particular piece of apparatus and to use it in a particular way carries an assumption that only certain sorts of circumstances will arise. There are instrumental as well as theoretical expectations, and they have often played a decisive role in scientific development. One such expectation is, for example, part of the story of oxygen's belated discovery. Using a standard test for "the goodness of air," both Priestley and Lavoisier mixed two volumes of their gas with one volume of nitric oxide, shook the mixture over water, and measured the volume of the gaseous residue. The previous experience from which this standard procedure had evolved assured them that with atmospheric air the residue

[8] Silvanus P. Thompson, *The Life of Sir William Thomson Baron Kelvin of Largs* (London, 1910), II, 1125.

would be one volume and that for any other gas (or for polluted air) it would be greater. In the oxygen experiments both found a residue close to one volume and identified the gas accordingly. Only much later and in part through an accident did Priestley renounce the standard procedure and try mixing nitric oxide with his gas in other proportions. He then found that with quadruple the volume of nitric oxide there was almost no residue at all. His commitment to the original test procedure—a procedure sanctioned by much previous experience—had been simultaneously a commitment to the non-existence of gases that could behave as oxygen did.[9]

Illustrations of this sort could be multiplied by reference, for example, to the belated identification of uranium fission. One reason why that nuclear reaction proved especially difficult to recognize was that men who knew what to expect when bombarding uranium chose chemical tests aimed mainly at elements from the upper end of the periodic table.[10] Ought we conclude from the frequency with which such instrumental commitments prove misleading that science should abandon standard tests and standard instruments? That would result in an inconceivable method of research. Paradigm procedures and applications are as necessary to science as paradigm laws and theories, and they have the same effects. Inevitably they restrict the phenomenological field accessible for scientific investigation at any

[9] Conant, *op. cit.*, pp. 18–20.

[10] K. K. Darrow, "Nuclear Fission," *Bell System Technical Journal,* XIX (1940), 267–89. Krypton, one of the two main fission products, seems not to have been identified by chemical means until after the reaction was well understood. Barium, the other product, was almost identified chemically at a late stage of the investigation because, as it happened, that element had to be added to the radioactive solution to precipitate the heavy element for which nuclear chemists were looking. Failure to separate that added barium from the radioactive product finally led, after the reaction had been repeatedly investigated for almost five years, to the following report: "As chemists we should be led by this research . . . to change all the names in the preceding [reaction] schema and thus write Ba, La, Ce instead of Ra, Ac, Th. But as 'nuclear chemists,' with close affiliations to physics, we cannot bring ourselves to this leap which would contradict all previous experience of nuclear physics. It may be that a series of strange accidents renders our results deceptive" (Otto Hahn and Fritz Strassman, "Über den Nachweis und das Verhalten der bei der Bestrahlung des Urans mittels Neutronen entstehended Erdalkalimetalle," *Die Naturwissenschaften,* XXVII [1939], 15).

given time. Recognizing that much, we may simultaneously see an essential sense in which a discovery like X-rays necessitates paradigm change—and therefore change in both procedures and expectations—for a special segment of the scientific community. As a result, we may also understand how the discovery of X-rays could seem to open a strange new world to many scientists and could thus participate so effectively in the crisis that led to twentieth-century physics.

Our final example of scientific discovery, that of the Leyden jar, belongs to a class that may be described as theory-induced. Initially, the term may seem paradoxical. Much that has been said so far suggests that discoveries predicted by theory in advance are parts of normal science and result in no *new sort* of fact. I have, for example, previously referred to the discoveries of new chemical elements during the second half of the nineteenth century as proceeding from normal science in that way. But not all theories are paradigm theories. Both during pre-paradigm periods and during the crises that lead to large-scale changes of paradigm, scientists usually develop many speculative and unarticulated theories that can themselves point the way to discovery. Often, however, that discovery is not quite the one anticipated by the speculative and tentative hypothesis. Only as experiment and tentative theory are together articulated to a match does the discovery emerge and the theory become a paradigm.

The discovery of the Leyden jar displays all these features as well as the others we have observed before. When it began, there was no single paradigm for electrical research. Instead, a number of theories, all derived from relatively accessible phenomena, were in competition. None of them succeeded in ordering the whole variety of electrical phenomena very well. That failure is the source of several of the anomalies that provide background for the discovery of the Leyden jar. One of the competing schools of electricians took electricity to be a fluid, and that conception led a number of men to attempt bottling the fluid by holding a water-filled glass vial in their hands and touching the water to a conductor suspended from an active electrostatic generator. On removing the jar from the machine and touching the water (or a conductor connected to it) with his free hand, each of these investigators experienced a severe shock. Those first experiments did not, however, provide electricians with the Leyden jar. That device emerged more slowly, and it is again impossible to say just when its discovery was completed. The initial attempts to store electrical fluid worked only because investigators held the vial in their hands while standing upon the ground. Electricians had still to learn that the jar required an outer as well as an inner conducting coating and that the fluid is not really stored in the jar at all. Somewhere in the course of the investigations that showed them this, and which introduced them to several other anomalous effects, the device that we call the Leyden jar emerged. Furthermore, the experiments that led to its emergence, many of them performed by Franklin, were also the ones that necessitated the drastic revision of the fluid theory and thus provided the first full paradigm for electricity.[11]

To a greater or lesser extent (corresponding to the continuum from the shocking to the anticipated result), the characteristics common to the three examples above are characteristic of all discoveries from which new sorts of phenomena emerge. Those characteristics include: the previous awareness of anomaly, the gradual and simultaneous emergence of both observational and conceptual recognition, and the consequent change of paradigm categories and procedures often accompanied by resistance. There is even evidence that these same characteristics are built into the nature of the perceptual process itself. In a psychological experiment that deserves to be far better known outside the trade, Bruner and Postman asked experimental subjects to identify on short and controlled exposure a series of playing cards. Many of the cards were normal, but some were made anoma-

[11] For various stages in the Leyden jar's evolution, see I. B. Cohen, *Franklin and Newton: An Inquiry into Speculative Newtonian Experimental Science and Franklin's Work in Electricity as an Example Thereof* (Philadelphia, 1956), pp. 385–86, 400–406, 452–67, 506–7. The last stage is described by Whittaker, *op. cit.*, pp. 50–52.

lous, e.g., a red six of spades and a black four of hearts. Each experimental run was constituted by the display of a single card to a single subject in a series of gradually increased exposures. After each exposure the subject was asked what he had seen, and the run was terminated by two successive correct identifications.[12]

Even on the shortest exposures many subjects identified most of the cards, and after a small increase all the subjects identified them all. For the normal cards these identifications were usually correct, but the anomalous cards were almost always identified, without apparent hesitation or puzzlement, as normal. The black four of hearts might, for example, be identified as the four of either spades or hearts. Without any awareness of trouble, it was immediately fitted to one of the conceptual categories prepared by prior experience. One would not even like to say that the subjects had seen something different from what they identified. With a further increase of exposure to the anomalous cards, subjects did begin to hesitate and to display awareness of anomaly. Exposed, for example, to the red six of spades, some would say: That's the six of spades, but there's something wrong with it—the black has a red border. Further increase of exposure resulted in still more hesitation and confusion until finally, and sometimes quite suddenly, most subjects would produce the correct identification without hesitation. Moreover, after doing this with two or three of the anomalous cards, they would have little further difficulty with the others. A few subjects, however, were never able to make the requisite adjustment of their categories. Even at forty times the average exposure required to recognize normal cards for what they were, more than 10 per cent of the anomalous cards were not correctly identified. And the subjects who then failed often experienced acute personal distress. One of them exclaimed: "I can't make the suit out, whatever it is. It didn't even look like a card that time. I don't know what color it is now or whether it's a spade or a heart. I'm

[12] J. S. Bruner and Leo Postman, "On the Perception of Incongruity: A Paradigm," *Journal of Personality*, XVIII (1949), 206–23.

not even sure now what a spade looks like. My God!"[13] In the next section we shall occasionally see scientists behaving this way too.

Either as a metaphor or because it reflects the nature of the mind, that psychological experiment provides a wonderfully simple and cogent schema for the process of scientific discovery. In science, as in the playing card experiment, novelty emerges only with difficulty, manifested by resistance, against a background provided by expectation. Initially, only the anticipated and usual are experienced even under circumstances where anomaly is later to be observed. Further acquaintance, however, does result in awareness of something wrong or does relate the effect to something that has gone wrong before. That awareness of anomaly opens a period in which conceptual categories are adjusted until the initially anomalous has become the anticipated. At this point the discovery has been completed. I have already urged that that process or one very much like it is involved in the emergence of all fundamental scientific novelties. Let me now point out that, recognizing the process, we can at last begin to see why normal science, a pursuit not directed to novelties and tending at first to suppress them, should nevertheless be so effective in causing them to arise.

In the development of any science, the first received paradigm is usually felt to account quite successfully for most of the observations and experiments easily accessible to that science's practitioners. Further development, therefore, ordinarily calls for the construction of elaborate equipment, the development of an esoteric vocabulary and skills, and a refinement of concepts that increasingly lessens their resemblance to their usual common-sense prototypes. That professionalization leads, on the one hand, to an immense restriction of the scientist's vision and to a considerable resistance to paradigm change. The science has become increasingly rigid. On the other hand, within those areas to which the paradigm directs the attention of the

[13] *Ibid.*, p. 218. My colleague Postman tells me that, though knowing all about the apparatus and display in advance, he nevertheless found looking at the incongruous cards acutely uncomfortable.

group, normal science leads to a detail of information and to a precision of the observation-theory match that could be achieved in no other way. Furthermore, that detail and precision-of-match have a value that transcends their not always very high intrinsic interest. Without the special apparatus that is constructed mainly for anticipated functions, the results that lead ultimately to novelty could not occur. And even when the apparatus exists, novelty ordinarily emerges only for the man who, knowing *with precision* what he should expect, is able to recognize that something has gone wrong. Anomaly appears only against the background provided by the paradigm. The more precise and far-reaching that paradigm is, the more sensitive an indicator it provides of anomaly and hence of an occasion for paradigm change. In the normal mode of discovery, even resistance to change has a use that will be explored more fully in the next section. By ensuring that the paradigm will not be too easily surrendered, resistance guarantees that scientists will not be lightly distracted and that the anomalies that lead to paradigm change will penetrate existing knowledge to the core. The very fact that a significant scientific novelty so often emerges simultaneously from several laboratories is an index both to the strongly traditional nature of normal science and to the completeness with which that traditional pursuit prepares the way for its own change.

VII. Crisis and the Emergence of Scientific Theories

All the discoveries considered in Section VI were causes of or contributors to paradigm change. Furthermore, the changes in which these discoveries were implicated were all destructive as well as constructive. After the discovery had been assimilated, scientists were able to account for a wider range of natural phenomena or to account with greater precision for some of those previously known. But that gain was achieved only by discarding some previously standard beliefs or procedures and, simultaneously, by replacing those components of the previous paradigm with others. Shifts of this sort are, I have argued, associated with all discoveries achieved through normal science, excepting only the unsurprising ones that had been anticipated in all but their details. Discoveries are not, however, the only sources of these destructive-constructive paradigm changes. In this section we shall begin to consider the similar, but usually far larger, shifts that result from the invention of new theories.

Having argued already that in the sciences fact and theory, discovery and invention, are not categorically and permanently distinct, we can anticipate overlap between this section and the last. (The impossible suggestion that Priestley first discovered oxygen and Lavoisier then invented it has its attractions. Oxygen has already been encountered as discovery; we shall shortly meet it again as invention.) In taking up the emergence of new theories we shall inevitably extend our understanding of discovery as well. Still, overlap is not identity. The sorts of discoveries considered in the last section were not, at least singly, responsible for such paradigm shifts as the Copernican, Newtonian, chemical, and Einsteinian revolutions. Nor were they responsible for the somewhat smaller, because more exclusively professional, changes in paradigm produced by the wave theory of light, the dynamical theory of heat, or Maxwell's electromagnetic theory. How can theories like these arise from normal science, an activity even less directed to their pursuit than to that of discoveries?

If awareness of anomaly plays a role in the emergence of new sorts of phenomena, it should surprise no one that a similar but more profound awareness is prerequisite to all acceptable changes of theory. On this point historical evidence is, I think, entirely unequivocal. The state of Ptolemaic astronomy was a scandal before Copernicus' announcement.[1] Galileo's contributions to the study of motion depended closely upon difficulties discovered in Aristotle's theory by scholastic critics.[2] Newton's new theory of light and color originated in the discovery that none of the existing pre-paradigm theories would account for the length of the spectrum, and the wave theory that replaced Newton's was announced in the midst of growing concern about anomalies in the relation of diffraction and polarization effects to Newton's theory.[3] Thermodynamics was born from the collision of two existing nineteenth-century physical theories, and quantum mechanics from a variety of difficulties surrounding black-body radiation, specific heats, and the photoelectric effect.[4] Furthermore, in all these cases except that of Newton the awareness of anomaly had lasted so long and penetrated so deep that one can appropriately describe the fields affected by it as in a state of growing crisis. Because it demands large-scale paradigm destruction and major shifts in the problems and techniques of normal science, the emergence of new theories is generally preceded by a period of pronounced professional in-

1 A. R. Hall, *The Scientific Revolution, 1500–1800* (London, 1954), p. 16.

2 Marshall Clagett, *The Science of Mechanics in the Middle Ages* (Madison, Wis., 1959), Parts II–III. A. Koyré displays a number of medieval elements in Galileo's thought in his *Etudes Galiléennes* (Paris, 1939), particularly Vol. I.

3 For Newton, see T. S. Kuhn, "Newton's Optical Papers," in *Isaac Newton's Papers and Letters in Natural Philosophy*, ed. I. B. Cohen (Cambridge, Mass., 1958), pp. 27–45. For the prelude to the wave theory, see E. T. Whittaker, *A History of the Theories of Aether and Electricity*, I (2d ed.; London, 1951), 94–109; and W. Whewell, *History of the Inductive Sciences* (rev. ed.; London, 1847), II, 396–466.

4 For thermodynamics, see Silvanus P. Thompson, *Life of William Thomson Baron Kelvin of Largs* (London, 1910), I, 266–81. For the quantum theory, see Fritz Reiche, *The Quantum Theory*, trans. H. S. Hatfield and H. L. Brose (London, 1922), chaps. i–ii.

security. As one might expect, that insecurity is generated by the persistent failure of the puzzles of normal science to come out as they should. Failure of existing rules is the prelude to a search for new ones.

Look first at a particularly famous case of paradigm change, the emergence of Copernican astronomy. When its predecessor, the Ptolemaic system, was first developed during the last two centuries before Christ and the first two after, it was admirably successful in predicting the changing positions of both stars and planets. No other ancient system had performed so well; for the stars, Ptolemaic astronomy is still widely used today as an engineering approximation; for the planets, Ptolemy's predictions were as good as Copernicus'. But to be admirably successful is never, for a scientific theory, to be completely successful. With respect both to planetary position and to precession of the equinoxes, predictions made with Ptolemy's system never quite conformed with the best available observations. Further reduction of those minor discrepancies constituted many of the principal problems of normal astronomical research for many of Ptolemy's successors, just as a similar attempt to bring celestial observation and Newtonian theory together provided normal research problems for Newton's eighteenth-century successors. For some time astronomers had every reason to suppose that these attempts would be as successful as those that had led to Ptolemy's system. Given a particular discrepancy, astronomers were invariably able to eliminate it by making some particular adjustment in Ptolemy's system of compounded circles. But as time went on, a man looking at the net result of the normal research effort of many astronomers could observe that astronomy's complexity was increasing far more rapidly than its accuracy and that a discrepancy corrected in one place was likely to show up in another.[5]

Because the astronomical tradition was repeatedly interrupted from outside and because, in the absence of printing, communication between astronomers was restricted, these dif-

[5] J. L. E. Dreyer, *A History of Astronomy from Thales to Kepler* (2d ed.; New York, 1953), chaps. xi–xii.

ficulties were only slowly recognized. But awareness did come. By the thirteenth century Alfonso X could proclaim that if God had consulted him when creating the universe, he would have received good advice. In the sixteenth century, Copernicus' coworker, Domenico da Novara, held that no system so cumbersome and inaccurate as the Ptolemaic had become could possibly be true of nature. And Copernicus himself wrote in the Preface to the *De Revolutionibus* that the astronomical tradition he inherited had finally created only a monster. By the early sixteenth century an increasing number of Europe's best astronomers were recognizing that the astronomical paradigm was failing in application to its own traditional problems. That recognition was prerequisite to Copernicus' rejection of the Ptolemaic paradigm and his search for a new one. His famous preface still provides one of the classic descriptions of a crisis state.[6]

Breakdown of the normal technical puzzle-solving activity is not, of course, the only ingredient of the astronomical crisis that faced Copernicus. An extended treatment would also discuss the social pressure for calendar reform, a pressure that made the puzzle of precession particularly urgent. In addition, a fuller account would consider medieval criticism of Aristotle, the rise of Renaissance Neoplatonism, and other significant historical elements besides. But technical breakdown would still remain the core of the crisis. In a mature science—and astronomy had become that in antiquity—external factors like those cited above are principally significant in determining the timing of breakdown, the ease with which it can be recognized, and the area in which, because it is given particular attention, the breakdown first occurs. Though immensely important, issues of that sort are out of bounds for this essay.

If that much is clear in the case of the Copernican revolution, let us turn from it to a second and rather different example, the crisis that preceded the emergence of Lavoisier's oxygen theory of combustion. In the 1770's many factors combined to generate

[6] T. S. Kuhn, *The Copernican Revolution* (Cambridge, Mass., 1957), pp. 135–43.

a crisis in chemistry, and historians are not altogether agreed about either their nature or their relative importance. But two of them are generally accepted as of first-rate significance: the rise of pneumatic chemistry and the question of weight relations. The history of the first begins in the seventeenth century with development of the air pump and its deployment in chemical experimentation. During the following century, using that pump and a number of other pneumatic devices, chemists came increasingly to realize that air must be an active ingredient in chemical reactions. But with a few exceptions—so equivocal that they may not be exceptions at all—chemists continued to believe that air was the only sort of gas. Until 1756, when Joseph Black showed that fixed air (CO_2) was consistently distinguishable from normal air, two samples of gas were thought to be distinct only in their impurities.[7]

After Black's work the investigation of gases proceeded rapidly, most notably in the hands of Cavendish, Priestley, and Scheele, who together developed a number of new techniques capable of distinguishing one sample of gas from another. All these men, from Black through Scheele, believed in the phlogiston theory and often employed it in their design and interpretation of experiments. Scheele actually first produced oxygen by an elaborate chain of experiments designed to dephlogisticate heat. Yet the net result of their experiments was a variety of gas samples and gas properties so elaborate that the phlogiston theory proved increasingly little able to cope with laboratory experience. Though none of these chemists suggested that the theory should be replaced, they were unable to apply it consistently. By the time Lavoisier began his experiments on airs in the early 1770's, there were almost as many versions of the phlogiston theory as there were pneumatic chemists.[8] That

[7] J. R. Partington, *A Short History of Chemistry* (2d ed.; London, 1951), pp. 48–51, 73–85, 90–120.

[8] Though their main concern is with a slightly later period, much relevant material is scattered throughout J. R. Partington and Douglas McKie's "Historical Studies on the Phlogiston Theory," *Annals of Science*, II (1937), 361–404; III (1938), 1–58, 337–71; and IV (1939), 337–71.

proliferation of versions of a theory is a very usual symptom of crisis. In his preface, Copernicus complained of it as well.

The increasing vagueness and decreasing utility of the phlogiston theory for pneumatic chemistry were not, however, the only source of the crisis that confronted Lavoisier. He was also much concerned to explain the gain in weight that most bodies experience when burned or roasted, and that again is a problem with a long prehistory. At least a few Islamic chemists had known that some metals gain weight when roasted. In the seventeenth century several investigators had concluded from this same fact that a roasted metal takes up some ingredient from the atmosphere. But in the seventeenth century that conclusion seemed unnecessary to most chemists. If chemical reactions could alter the volume, color, and texture of the ingredients, why should they not alter weight as well? Weight was not always taken to be the measure of quantity of matter. Besides, weight-gain on roasting remained an isolated phenomenon. Most natural bodies (e.g., wood) lose weight on roasting as the phlogiston theory was later to say they should.

During the eighteenth century, however, these initially adequate responses to the problem of weight-gain became increasingly difficult to maintain. Partly because the balance was increasingly used as a standard chemical tool and partly because the development of pneumatic chemistry made it possible and desirable to retain the gaseous products of reactions, chemists discovered more and more cases in which weight-gain accompanied roasting. Simultaneously, the gradual assimilation of Newton's gravitational theory led chemists to insist that gain in weight must mean gain in quantity of matter. Those conclusions did not result in rejection of the phlogiston theory, for that theory could be adjusted in many ways. Perhaps phlogiston had negative weight, or perhaps fire particles or something else entered the roasted body as phlogiston left it. There were other explanations besides. But if the problem of weight-gain did not lead to rejection, it did lead to an increasing number of special studies in which this problem bulked large. One of them, "On phlogiston considered as a substance with weight and [analyzed] in terms of the weight changes it produces in bodies with which it unites," was read to the French Academy early in 1772, the year which closed with Lavoisier's delivery of his famous sealed note to the Academy's Secretary. Before that note was written a problem that had been at the edge of the chemist's consciousness for many years had become an outstanding unsolved puzzle.[9] Many different versions of the phlogiston theory were being elaborated to meet it. Like the problems of pneumatic chemistry, those of weight-gain were making it harder and harder to know what the phlogiston theory was. Though still believed and trusted as a working tool, a paradigm of eighteenth-century chemistry was gradually losing its unique status. Increasingly, the research it guided resembled that conducted under the competing schools of the pre-paradigm period, another typical effect of crisis.

Consider now, as a third and final example, the late nineteenth century crisis in physics that prepared the way for the emergence of relativity theory. One root of that crisis can be traced to the late seventeenth century when a number of natural philosophers, most notably Leibniz, criticized Newton's retention of an updated version of the classic conception of absolute space.[10] They were very nearly, though never quite, able to show that absolute positions and absolute motions were without any function at all in Newton's system; and they did succeed in hinting at the considerable aesthetic appeal a fully relativistic conception of space and motion would later come to display. But their critique was purely logical. Like the early Copernicans who criticized Aristotle's proofs of the earth's stability, they did not dream that transition to a relativistic system could have observational consequences. At no point did they relate their views to any problems that arose when applying Newtonian theory to nature. As a result, their views died with

[9] H. Guerlac, *Lavoisier—the Crucial Year* (Ithaca, N.Y., 1961). The entire book documents the evolution and first recognition of a crisis. For a clear statement of the situation with respect to Lavoisier, see p. 35.

[10] Max Jammer, *Concepts of Space: The History of Theories of Space in Physics* (Cambridge, Mass., 1954), pp. 114–24.

them during the early decades of the eighteenth century to be resurrected only in the last decades of the nineteenth when they had a very different relation to the practice of physics.

The technical problems to which a relativistic philosophy of space was ultimately to be related began to enter normal science with the acceptance of the wave theory of light after about 1815, though they evoked no crisis until the 1890's. If light is wave motion propagated in a mechanical ether governed by Newton's Laws, then both celestial observation and terrestrial experiment become potentially capable of detecting drift through the ether. Of the celestial observations, only those of aberration promised sufficient accuracy to provide relevant information, and the detection of ether-drift by aberration measurements therefore became a recognized problem for normal research. Much special equipment was built to resolve it. That equipment, however, detected no observable drift, and the problem was therefore transferred from the experimentalists and observers to the theoreticians. During the central decades of the century Fresnel, Stokes, and others devised numerous articulations of the ether theory designed to explain the failure to observe drift. Each of these articulations assumed that a moving body drags some fraction of the ether with it. And each was sufficiently successful to explain the negative results not only of celestial observation but also of terrestrial experimentation, including the famous experiment of Michelson and Morley.[11] There was still no conflict excepting that between the various articulations. In the absence of relevant experimental techniques, that conflict never became acute.

The situation changed again only with the gradual acceptance of Maxwell's electromagnetic theory in the last two decades of the nineteenth century. Maxwell himself was a Newtonian who believed that light and electromagnetism in general were due to variable displacements of the particles of a mechanical ether. His earliest versions of a theory for electricity and

[11] Joseph Larmor, *Aether and Matter . . . Including a Discussion of the Influence of the Earth's Motion on Optical Phenomena* (Cambridge, 1900), pp. 6–20, 320–22.

magnetism made direct use of hypothetical properties with which he endowed this medium. These were dropped from his final version, but he still believed his electromagnetic theory compatible with some articulation of the Newtonian mechanical view.[12] Developing a suitable articulation was a challenge for him and his successors. In practice, however, as has happened again and again in scientific development, the required articulation proved immensely difficult to produce. Just as Copernicus' astronomical proposal, despite the optimism of its author, created an increasing crisis for existing theories of motion, so Maxwell's theory, despite its Newtonian origin, ultimately produced a crisis for the paradigm from which it had sprung.[13] Furthermore, the locus at which that crisis became most acute was provided by the problems we have just been considering, those of motion with respect to the ether.

Maxwell's discussion of the electromagnetic behavior of bodies in motion had made no reference to ether drag, and it proved very difficult to introduce such drag into his theory. As a result, a whole series of earlier observations designed to detect drift through the ether became anomalous. The years after 1890 therefore witnessed a long series of attempts, both experimental and theoretical, to detect motion with respect to the ether and to work ether drag into Maxwell's theory. The former were uniformly unsuccessful, though some analysts thought their results equivocal. The latter produced a number of promising starts, particularly those of Lorentz and Fitzgerald, but they also disclosed still other puzzles and finally resulted in just that proliferation of competing theories that we have previously found to be the concomitant of crisis.[14] It is against that historical setting that Einstein's special theory of relativity emerged in 1905.

These three examples are almost entirely typical. In each case a novel theory emerged only after a pronounced failure in the

[12] R. T. Glazebrook, *James Clerk Maxwell and Modern Physics* (London, 1896), chap. ix. For Maxwell's final attitude, see his own book, *A Treatise on Electricity and Magnetism* (3d ed.; Oxford, 1892), p. 470.

[13] For astronomy's role in the development of mechanics, see Kuhn, *op. cit.*, chap. vii.

[14] Whittaker, *op. cit.*, I, 386–410; and II (London, 1953), 27–40.

normal problem-solving activity. Furthermore, except for the case of Copernicus in which factors external to science played a particularly large role, that breakdown and the proliferation of theories that is its sign occurred no more than a decade or two before the new theory's enunciation. The novel theory seems a direct response to crisis. Note also, though this may not be quite so typical, that the problems with respect to which breakdown occurred were all of a type that had long been recognized. Previous practice of normal science had given every reason to consider them solved or all but solved, which helps to explain why the sense of failure, when it came, could be so acute. Failure with a new sort of problem is often disappointing but never surprising. Neither problems nor puzzles yield often to the first attack. Finally, these examples share another characteristic that may help to make the case for the role of crisis impressive: the solution to each of them had been at least partially anticipated during a period when there was no crisis in the corresponding science; and in the absence of crisis those anticipations had been ignored.

The only complete anticipation is also the most famous, that of Copernicus by Aristarchus in the third century B.C. It is often said that if Greek science had been less deductive and less ridden by dogma, heliocentric astronomy might have begun its development eighteen centuries earlier than it did.[15] But that is to ignore all historical context. When Aristarchus' suggestion was made, the vastly more reasonable geocentric system had no needs that a heliocentric system might even conceivably have fulfilled. The whole development of Ptolemaic astronomy, both its triumphs and its breakdown, falls in the centuries after Aristarchus' proposal. Besides, there were no obvious reasons for taking Aristarchus seriously. Even Copernicus' more elaborate proposal was neither simpler nor more accurate than Ptolemy's system. Available observational tests, as we shall see more clear-

15 For Aristarchus' work, see T. L. Heath, *Aristarchus of Samos: The Ancient Copernicus* (Oxford, 1913), Part II. For an extreme statement of the traditional position about the neglect of Aristarchus' achievement, see Arthur Koestler, *The Sleepwalkers: A History of Man's Changing Vision of the Universe* (London, 1959), p. 50.

ly below, provided no basis for a choice between them. Under those circumstances, one of the factors that led astronomers to Copernicus (and one that could not have led them to Aristarchus) was the recognized crisis that had been responsible for innovation in the first place. Ptolemaic astronomy had failed to solve its problems; the time had come to give a competitor a chance. Our other two examples provide no similarly full anticipations. But surely one reason why the theories of combustion by absorption from the atmosphere—theories developed in the seventeenth century by Rey, Hooke, and Mayow—failed to get a sufficient hearing was that they made no contact with a recognized trouble spot in normal scientific practice.[16] And the long neglect by eighteenth- and nineteenth-century scientists of Newton's relativistic critics must largely have been due to a similar failure in confrontation.

Philosophers of science have repeatedly demonstrated that more than one theoretical construction can always be placed upon a given collection of data. History of science indicates that, particularly in the early developmental stages of a new paradigm, it is not even very difficult to invent such alternates. But that invention of alternates is just what scientists seldom undertake except during the pre-paradigm stage of their science's development and at very special occasions during its subsequent evolution. So long as the tools a paradigm supplies continue to prove capable of solving the problems it defines, science moves fastest and penetrates most deeply through confident employment of those tools. The reason is clear. As in manufacture so in science—retooling is an extravagance to be reserved for the occasion that demands it. The significance of crises is the indication they provide that an occasion for retooling has arrived.

16 Partington, *op. cit.*, pp. 78–85.

VIII. The Response to Crisis

Let us then assume that crises are a necessary precondition for the emergence of novel theories and ask next how scientists respond to their existence. Part of the answer, as obvious as it is important, can be discovered by noting first what scientists never do when confronted by even severe and prolonged anomalies. Though they may begin to lose faith and then to consider alternatives, they do not renounce the paradigm that has led them into crisis. They do not, that is, treat anomalies as counter-instances, though in the vocabulary of philosophy of science that is what they are. In part this generalization is simply a statement from historic fact, based upon examples like those given above and, more extensively, below. These hint what our later examination of paradigm rejection will disclose more fully: once it has achieved the status of paradigm, a scientific theory is declared invalid only if an alternate candidate is available to take its place. No process yet disclosed by the historical study of scientific development at all resembles the methodological stereotype of falsification by direct comparison with nature. That remark does not mean that scientists do not reject scientific theories, or that experience and experiment are not essential to the process in which they do so. But it does mean—what will ultimately be a central point—that the act of judgment that leads scientists to reject a previously accepted theory is always based upon more than a comparison of that theory with the world. The decision to reject one paradigm is always simultaneously the decision to accept another, and the judgment leading to that decision involves the comparison of both paradigms with nature *and* with each other.

There is, in addition, a second reason for doubting that scientists reject paradigms because confronted with anomalies or counterinstances. In developing it my argument will itself foreshadow another of this essay's main theses. The reasons for doubt sketched above were purely factual; they were, that is, themselves counterinstances to a prevalent epistemological theory. As such, if my present point is correct, they can at best help to create a crisis or, more accurately, to reinforce one that is already very much in existence. By themselves they cannot and will not falsify that philosophical theory, for its defenders will do what we have already seen scientists doing when confronted by anomaly. They will devise numerous articulations and *ad hoc* modifications of their theory in order to eliminate any apparent conflict. Many of the relevant modifications and qualifications are, in fact, already in the literature. If, therefore, these epistemological counterinstances are to constitute more than a minor irritant, that will be because they help to permit the emergence of a new and different analysis of science within which they are no longer a source of trouble. Furthermore, if a typical pattern, which we shall later observe in scientific revolutions, is applicable here, these anomalies will then no longer seem to be simply facts. From within a new theory of scientific knowledge, they may instead seem very much like tautologies, statements of situations that could not conceivably have been otherwise.

It has often been observed, for example, that Newton's second law of motion, though it took centuries of difficult factual and theoretical research to achieve, behaves for those committed to Newton's theory very much like a purely logical statement that no amount of observation could refute.[1] In Section X we shall see that the chemical law of fixed proportion, which before Dalton was an occasional experimental finding of very dubious generality, became after Dalton's work an ingredient of a definition of chemical compound that no experimental work could by itself have upset. Something much like that will also happen to the generalization that scientists fail to reject paradigms when faced with anomalies or counterinstances. They could not do so and still remain scientists.

Though history is unlikely to record their names, some men have undoubtedly been driven to desert science because of

[1] See particularly the discussion in N. R. Hanson, *Patterns of Discovery* (Cambridge, 1958), pp. 99–105.

their inability to tolerate crisis. Like artists, creative scientists must occasionally be able to live in a world out of joint—elsewhere I have described that necessity as "the essential tension" implicit in scientific research.[2] But that rejection of science in favor of another occupation is, I think, the only sort of paradigm rejection to which counterinstances by themselves can lead. Once a first paradigm through which to view nature has been found, there is no such thing as research in the absence of any paradigm. To reject one paradigm without simultaneously substituting another is to reject science itself. That act reflects not on the paradigm but on the man. Inevitably he will be seen by his colleagues as "the carpenter who blames his tools."

The same point can be made at least equally effectively in reverse: there is no such thing as research without counterinstances. For what is it that differentiates normal science from science in a crisis state? Not, surely, that the former confronts no counterinstances. On the contrary, what we previously called the puzzles that constitute normal science exist only because no paradigm that provides a basis for scientific research ever completely resolves all its problems. The very few that have ever seemed to do so (e.g., geometric optics) have shortly ceased to yield research problems at all and have instead become tools for engineering. Excepting those that are exclusively instrumental, every problem that normal science sees as a puzzle can be seen, from another viewpoint, as a counterinstance and thus as a source of crisis. Copernicus saw as counterinstances what most of Ptolemy's other successors had seen as puzzles in the match between observation and theory. Lavoisier saw as a counterinstance what Priestley had seen as a successfully solved puzzle in the articulation of the phlogiston theory. And Einstein saw as counterinstances what Lorentz, Fitzgerald, and others had seen as puzzles in the articulation of Newton's and Max-

[2] T. S. Kuhn, "The Essential Tension: Tradition and Innovation in Scientific Research," in *The Third (1959) University of Utah Research Conference on the Identification of Creative Scientific Talent,* ed. Calvin W. Taylor (Salt Lake City, 1959), pp. 162–77. For the comparable phenomenon among artists, see Frank Barron, "The Psychology of Imagination," *Scientific American,* CXCIX (September, 1958), 151–66, esp. 160.

well's theories. Furthermore, even the existence of crisis does not by itself transform a puzzle into a counterinstance. There is no such sharp dividing line. Instead, by proliferating versions of the paradigm, crisis loosens the rules of normal puzzle-solving in ways that ultimately permit a new paradigm to emerge. There are, I think, only two alternatives: either no scientific theory ever confronts a counterinstance, or all such theories confront counterinstances at all times.

How can the situation have seemed otherwise? That question necessarily leads to the historical and critical elucidation of philosophy, and those topics are here barred. But we can at least note two reasons why science has seemed to provide so apt an illustration of the generalization that truth and falsity are uniquely and unequivocally determined by the confrontation of statement with fact. Normal science does and must continually strive to bring theory and fact into closer agreement, and that activity can easily be seen as testing or as a search for confirmation or falsification. Instead, its object is to solve a puzzle for whose very existence the validity of the paradigm must be assumed. Failure to achieve a solution discredits only the scientist and not the theory. Here, even more than above, the proverb applies: "It is a poor carpenter who blames his tools." In addition, the manner in which science pedagogy entangles discussion of a theory with remarks on its exemplary applications has helped to reinforce a confirmation-theory drawn predominantly from other sources. Given the slightest reason for doing so, the man who reads a science text can easily take the applications to be the evidence for the theory, the reasons why it ought to be believed. But science students accept theories on the authority of teacher and text, not because of evidence. What alternatives have they, or what competence? The applications given in texts are not there as evidence but because learning them is part of learning the paradigm at the base of current practice. If applications were set forth as evidence, then the very failure of texts to suggest alternative interpretations or to discuss problems for which scientists have failed to produce paradigm solutions would convict their authors of extreme bias. There is not the slightest reason for such an indictment.

How, then, to return to the initial question, do scientists respond to the awareness of an anomaly in the fit between theory and nature? What has just been said indicates that even a discrepancy unaccountably larger than that experienced in other applications of the theory need not draw any very profound response. There are always some discrepancies. Even the most stubborn ones usually respond at last to normal practice. Very often scientists are willing to wait, particularly if there are many problems available in other parts of the field. We have already noted, for example, that during the sixty years after Newton's original computation, the predicted motion of the moon's perigee remained only half of that observed. As Europe's best mathematical physicists continued to wrestle unsuccessfully with the well-known discrepancy, there were occasional proposals for a modification of Newton's inverse square law. But no one took these proposals very seriously, and in practice this patience with a major anomaly proved justified. Clairaut in 1750 was able to show that only the mathematics of the application had been wrong and that Newtonian theory could stand as before.[3] Even in cases where no mere mistake seems quite possible (perhaps because the mathematics involved is simpler or of a familiar and elsewhere successful sort), persistent and recognized anomaly does not always induce crisis. No one seriously questioned Newtonian theory because of the long-recognized discrepancies between predictions from that theory and both the speed of sound and the motion of Mercury. The first discrepancy was ultimately and quite unexpectedly resolved by experiments on heat undertaken for a very different purpose; the second vanished with the general theory of relativity after a crisis that it had had no role in creating.[4] Apparent-

[3] W. Whewell, *History of the Inductive Sciences* (rev. ed.; London, 1847), II, 220–21.

[4] For the speed of sound, see T. S. Kuhn, "The Caloric Theory of Adiabatic Compression," *Isis,* XLIV (1958), 136–37. For the secular shift in Mercury's perihelion, see E. T. Whittaker, *A History of the Theories of Aether and Electricity,* II (London, 1953), 151, 179.

ly neither had seemed sufficiently fundamental to evoke the malaise that goes with crisis. They could be recognized as counterinstances and still be set aside for later work.

It follows that if an anomaly is to evoke crisis, it must usually be more than just an anomaly. There are always difficulties somewhere in the paradigm-nature fit; most of them are set right sooner or later, often by processes that could not have been foreseen. The scientist who pauses to examine every anomaly he notes will seldom get significant work done. We therefore have to ask what it is that makes an anomaly seem worth concerted scrutiny, and to that question there is probably no fully general answer. The cases we have already examined are characteristic but scarcely prescriptive. Sometimes an anomaly will clearly call into question explicit and fundamental generalizations of the paradigm, as the problem of ether drag did for those who accepted Maxwell's theory. Or, as in the Copernican revolution, an anomaly without apparent fundamental import may evoke crisis if the applications that it inhibits have a particular practical importance, in this case for calendar design and astrology. Or, as in eighteenth-century chemistry, the development of normal science may transform an anomaly that had previously been only a vexation into a source of crisis: the problem of weight relations had a very different status after the evolution of pneumatic-chemical techniques. Presumably there are still other circumstances that can make an anomaly particularly pressing, and ordinarily several of these will combine. We have already noted, for example, that one source of the crisis that confronted Copernicus was the mere length of time during which astronomers had wrestled unsuccessfully with the reduction of the residual discrepancies in Ptolemy's system.

When, for these reasons or others like them, an anomaly comes to seem more than just another puzzle of normal science, the transition to crisis and to extraordinary science has begun. The anomaly itself now comes to be more generally recognized as such by the profession. More and more attention is devoted to it by more and more of the field's most eminent men. If it still continues to resist, as it usually does not, many of them may come to view its resolution as *the* subject matter of their discipline. For them the field will no longer look quite the same as it had earlier. Part of its different appearance results simply from the new fixation point of scientific scrutiny. An even more important source of change is the divergent nature of the numerous partial solutions that concerted attention to the problem has made available. The early attacks upon the resistant problem will have followed the paradigm rules quite closely. But with continuing resistance, more and more of the attacks upon it will have involved some minor or not so minor articulation of the paradigm, no two of them quite alike, each partially successful, but none sufficiently so to be accepted as paradigm by the group. Through this proliferation of divergent articulations (more and more frequently they will come to be described as *ad hoc* adjustments), the rules of normal science become increasingly blurred. Though there still is a paradigm, few practitioners prove to be entirely agreed about what it is. Even formerly standard solutions of solved problems are called in question.

When acute, this situation is sometimes recognized by the scientists involved. Copernicus complained that in his day astronomers were so "inconsistent in these [astronomical] investigations . . . that they cannot even explain or observe the constant length of the seasonal year." "With them," he continued, "it is as though an artist were to gather the hands, feet, head and other members for his images from diverse models, each part excellently drawn, but not related to a single body, and since they in no way match each other, the result would be monster rather than man."[5] Einstein, restricted by current usage to less florid language, wrote only, "It was as if the ground had been pulled out from under one, with no firm foundation to be seen anywhere, upon which one could have built."[6] And Wolfgang Pauli, in the months before Heisenberg's paper on matrix

[5] Quoted in T. S. Kuhn, *The Copernican Revolution* (Cambridge, Mass., 1957), p. 138.

[6] Albert Einstein, "Autobiographical Note," in *Albert Einstein: Philosopher-Scientist,* ed. P. A. Schilpp (Evanston, Ill., 1949), p. 45.

mechanics pointed the way to a new quantum theory, wrote to a friend, "At the moment physics is again terribly confused. In any case, it is too difficult for me, and I wish I had been a movie comedian or something of the sort and had never heard of physics." That testimony is particularly impressive if contrasted with Pauli's words less than five months later: "Heisenberg's type of mechanics has again given me hope and joy in life. To be sure it does not supply the solution to the riddle, but I believe it is again possible to march forward."[7]

Such explicit recognitions of breakdown are extremely rare, but the effects of crisis do not entirely depend upon its conscious recognition. What can we say these effects are? Only two of them seem to be universal. All crises begin with the blurring of a paradigm and the consequent loosening of the rules for normal research. In this respect research during crisis very much resembles research during the pre-paradigm period, except that in the former the locus of difference is both smaller and more clearly defined. And all crises close in one of three ways. Sometimes normal science ultimately proves able to handle the crisis-provoking problem despite the despair of those who have seen it as the end of an existing paradigm. On other occasions the problem resists even apparently radical new approaches. Then scientists may conclude that no solution will be forthcoming in the present state of their field. The problem is labelled and set aside for a future generation with more developed tools. Or, finally, the case that will most concern us here, a crisis may end with the emergence of a new candidate for paradigm and with the ensuing battle over its acceptance. This last mode of closure will be considered at length in later sections, but we must anticipate a bit of what will be said there in order to complete these remarks about the evolution and anatomy of the crisis state.

The transition from a paradigm in crisis to a new one from which a new tradition of normal science can emerge is far from a cumulative process, one achieved by an articulation or exten-

[7] Ralph Kronig, "The Turning Point," in *Theoretical Physics in the Twentieth Century: A Memorial Volume to Wolfgang Pauli*, ed. M. Fierz and V. F. Weisskopf (New York, 1960), pp. 22, 25–26. Much of this article describes the crisis in quantum mechanics in the years immediately before 1925.

sion of the old paradigm. Rather it is a reconstruction of the field from new fundamentals, a reconstruction that changes some of the field's most elementary theoretical generalizations as well as many of its paradigm methods and applications. During the transition period there will be a large but never complete overlap between the problems that can be solved by the old and by the new paradigm. But there will also be a decisive difference in the modes of solution. When the transition is complete, the profession will have changed its view of the field, its methods, and its goals. One perceptive historian, viewing a classic case of a science's reorientation by paradigm change, recently described it as "picking up the other end of the stick," a process that involves "handling the same bundle of data as before, but placing them in a new system of relations with one another by giving them a different framework."[8] Others who have noted this aspect of scientific advance have emphasized its similarity to a change in visual gestalt: the marks on paper that were first seen as a bird are now seen as an antelope, or vice versa.[9] That parallel can be misleading. Scientists do not see something *as* something else; instead, they simply see it. We have already examined some of the problems created by saying that Priestley saw oxygen as dephlogisticated air. In addition, the scientist does not preserve the gestalt subject's freedom to switch back and forth between ways of seeing. Nevertheless, the switch of gestalt, particularly because it is today so familiar, is a useful elementary prototype for what occurs in full-scale paradigm shift.

The preceding anticipation may help us recognize crisis as an appropriate prelude to the emergence of new theories, particularly since we have already examined a small-scale version of the same process in discussing the emergence of discoveries. Just because the emergence of a new theory breaks with one tradition of scientific practice and introduces a new one conducted under different rules and within a different universe of

[8] Herbert Butterfield, *The Origins of Modern Science, 1300–1800* (London, 1949), pp. 1–7.

[9] Hanson, *op. cit.*, chap. i.

discourse, it is likely to occur only when the first tradition is felt to have gone badly astray. That remark is, however, no more than a prelude to the investigation of the crisis-state, and, unfortunately, the questions to which it leads demand the competence of the psychologist even more than that of the historian. What is extraordinary research like? How is anomaly made lawlike? How do scientists proceed when aware only that something has gone fundamentally wrong at a level with which their training has not equipped them to deal? Those questions need far more investigation, and it ought not all be historical. What follows will necessarily be more tentative and less complete than what has gone before.

Often a new paradigm emerges, at least in embryo, before a crisis has developed far or been explicitly recognized. Lavoisier's work provides a case in point. His sealed note was deposited with the French Academy less than a year after the first thorough study of weight relations in the phlogiston theory and before Priestley's publications had revealed the full extent of the crisis in pneumatic chemistry. Or again, Thomas Young's first accounts of the wave theory of light appeared at a very early stage of a developing crisis in optics, one that would be almost unnoticeable except that, with no assistance from Young, it had grown to an international scientific scandal within a decade of the time he first wrote. In cases like these one can say only that a minor breakdown of the paradigm and the very first blurring of its rules for normal science were sufficient to induce in someone a new way of looking at the field. What intervened between the first sense of trouble and the recognition of an available alternate must have been largely unconscious.

In other cases, however—those of Copernicus, Einstein, and contemporary nuclear theory, for example—considerable time elapses between the first consciousness of breakdown and the emergence of a new paradigm. When that occurs, the historian may capture at least a few hints of what extraordinary science is like. Faced with an admittedly fundamental anomaly in theory, the scientist's first effort will often be to isolate it more precisely and to give it structure. Though now aware that they cannot be quite right, he will push the rules of normal science harder than ever to see, in the area of difficulty, just where and how far they can be made to work. Simultaneously he will seek for ways of magnifying the breakdown, of making it more striking and perhaps also more suggestive than it had been when displayed in experiments the outcome of which was thought to be known in advance. And in the latter effort, more than in any other part of the post-paradigm development of science, he will look almost like our most prevalent image of the scientist. He will, in the first place, often seem a man searching at random, trying experiments just to see what will happen, looking for an effect whose nature he cannot quite guess. Simultaneously, since no experiment can be conceived without some sort of theory, the scientist in crisis will constantly try to generate speculative theories that, if successful, may disclose the road to a new paradigm and, if unsuccessful, can be surrendered with relative ease.

Kepler's account of his prolonged struggle with the motion of Mars and Priestley's description of his response to the proliferation of new gases provide classic examples of the more random sort of research produced by the awareness of anomaly.[10] But probably the best illustrations of all come from contemporary research in field theory and on fundamental particles. In the absence of a crisis that made it necessary to see just how far the rules of normal science could stretch, would the immense effort required to detect the neutrino have seemed justified? Or, if the rules had not obviously broken down at some undisclosed point, would the radical hypothesis of parity non-conservation have been either suggested or tested? Like much other research in physics during the past decade, these experiments were in part attempts to localize and define the source of a still diffuse set of anomalies.

This sort of extraordinary research is often, though by no

[10] For an account of Kepler's work on Mars, see J. L. E. Dreyer, *A History of Astronomy from Thales to Kepler* (2d ed.; New York, 1953), pp. 380–93. Occasional inaccuracies do not prevent Dreyer's précis from providing the material needed here. For Priestley, see his own work, esp. *Experiments and Observations on Different Kinds of Air* (London, 1774–75).

means generally, accompanied by another. It is, I think, particularly in periods of acknowledged crisis that scientists have turned to philosophical analysis as a device for unlocking the riddles of their field. Scientists have not generally needed or wanted to be philosophers. Indeed, normal science usually holds creative philosophy at arm's length, and probably for good reasons. To the extent that normal research work can be conducted by using the paradigm as a model, rules and assumptions need not be made explicit. In Section V we noted that the full set of rules sought by philosophical analysis need not even exist. But that is not to say that the search for assumptions (even for non-existent ones) cannot be an effective way to weaken the grip of a tradition upon the mind and to suggest the basis for a new one. It is no accident that the emergence of Newtonian physics in the seventeenth century and of relativity and quantum mechanics in the twentieth should have been both preceded and accompanied by fundamental philosophical analyses of the contemporary research tradition.[11] Nor is it an accident that in both these periods the so-called thought experiment should have played so critical a role in the progress of research. As I have shown elsewhere, the analytical thought experimentation that bulks so large in the writings of Galileo, Einstein, Bohr, and others is perfectly calculated to expose the old paradigm to existing knowledge in ways that isolate the root of crisis with a clarity unattainable in the laboratory.[12]

With the deployment, singly or together, of these extraordinary procedures, one other thing may occur. By concentrating scientific attention upon a narrow area of trouble and by preparing the scientific mind to recognize experimental anomalies for what they are, crisis often proliferates new discoveries. We have already noted how the awareness of crisis distinguishes

[11] For the philosophical counterpoint that accompanied seventeenth-century mechanics, see René Dugas, *La mécanique au XVIIe siècle* (Neuchatel, 1954), particularly chap. xi. For the similar nineteenth-century episode, see the same author's earlier book, *Histoire de la mécanique* (Neuchatel, 1950), pp. 419–43.

[12] T. S. Kuhn, "A Function for Thought Experiments," in *Mélanges Alexandre Koyré*, ed. R. Taton and I. B. Cohen, to be published by Hermann (Paris) in 1963.

Lavoisier's work on oxygen from Priestley's; and oxygen was not the only new gas that the chemists aware of anomaly were able to discover in Priestley's work. Or again, new optical discoveries accumulated rapidly just before and during the emergence of the wave theory of light. Some, like polarization by reflection, were a result of the accidents that concentrated work in an area of trouble makes likely. (Malus, who made the discovery, was just starting work for the Academy's prize essay on double refraction, a subject widely known to be in an unsatisfactory state.) Others, like the light spot at the center of the shadow of a circular disk, were predictions from the new hypothesis, ones whose success helped to transform it to a paradigm for later work. And still others, like the colors of scratches and of thick plates, were effects that had often been seen and occasionally remarked before, but that, like Priestley's oxygen, had been assimilated to well-known effects in ways that prevented their being seen for what they were.[13] A similar account could be given of the multiple discoveries that, from about 1895, were a constant concomitant of the emergence of quantum mechanics.

Extraordinary research must have still other manifestations and effects, but in this area we have scarcely begun to discover the questions that need to be asked. Perhaps, however, no more are needed at this point. The preceding remarks should suffice to show how crisis simultaneously loosens the stereotypes and provides the incremental data necessary for a fundamental paradigm shift. Sometimes the shape of the new paradigm is foreshadowed in the structure that extraordinary research has given to the anomaly. Einstein wrote that before he had any substitute for classical mechanics, he could see the interrelation between the known anomalies of black-body radiation, the photoelectric effect, and specific heats.[14] More often no such structure is consciously seen in advance. Instead, the new paradigm, or a sufficient hint to permit later articulation, emerges

[13] For the new optical discoveries in general, see V. Ronchi, *Histoire de la lumière* (Paris, 1956), chap. vii. For the earlier explanation of one of these effects, see J. Priestley, *The History and Present State of Discoveries Relating to Vision, Light and Colours* (London, 1772), pp. 498–520.

[14] Einstein, *loc. cit.*

all at once, sometimes in the middle of the night, in the mind of a man deeply immersed in crisis. What the nature of that final stage is—how an individual invents (or finds he has invented) a new way of giving order to data now all assembled—must here remain inscrutable and may be permanently so. Let us here note only one thing about it. Almost always the men who achieve these fundamental inventions of a new paradigm have been either very young or very new to the field whose paradigm they change.[15] And perhaps that point need not have been made explicit, for obviously these are the men who, being little committed by prior practice to the traditional rules of normal science, are particularly likely to see that those rules no longer define a playable game and to conceive another set that can replace them.

The resulting transition to a new paradigm is scientific revolution, a subject that we are at long last prepared to approach directly. Note first, however, one last and apparently elusive respect in which the material of the last three sections has prepared the way. Until Section VI, where the concept of anomaly was first introduced, the terms 'revolution' and 'extraordinary science' may have seemed equivalent. More important, neither term may have seemed to mean more than 'non-normal science,' a circularity that will have bothered at least a few readers. In practice, it need not have done so. We are about to discover that a similar circularity is characteristic of scientific theories. Bothersome or not, however, that circularity is no longer unqualified. This section of the essay and the two preceding have educed numerous criteria of a breakdown in normal scientific activity, criteria that do not at all depend upon whether breakdown is succeeded by revolution. Confronted with anomaly or

[15] This generalization about the role of youth in fundamental scientific research is so common as to be a cliché. Furthermore, a glance at almost any list of fundamental contributions to scientific theory will provide impressionistic confirmation. Nevertheless, the generalization badly needs systematic investigation. Harvey C. Lehman (*Age and Achievement* [Princeton, 1953]) provides many useful data; but his studies make no attempt to single out contributions that involve fundamental reconceptualization. Nor do they inquire about the special circumstances, if any, that may accompany relatively late productivity in the sciences.

with crisis, scientists take a different attitude toward existing paradigms, and the nature of their research changes accordingly. The proliferation of competing articulations, the willingness to try anything, the expression of explicit discontent, the recourse to philosophy and to debate over fundamentals, all these are symptoms of a transition from normal to extraordinary research. It is upon their existence more than upon that of revolutions that the notion of normal science depends.

IX. The Nature and Necessity of Scientific Revolutions

These remarks permit us at last to consider the problems that provide this essay with its title. What are scientific revolutions, and what is their function in scientific development? Much of the answer to these questions has been anticipated in earlier sections. In particular, the preceding discussion has indicated that scientific revolutions are here taken to be those non-cumulative developmental episodes in which an older paradigm is replaced in whole or in part by an incompatible new one. There is more to be said, however, and an essential part of it can be introduced by asking one further question. Why should a change of paradigm be called a revolution? In the face of the vast and essential differences between political and scientific development, what parallelism can justify the metaphor that finds revolutions in both?

One aspect of the parallelism must already be apparent. Political revolutions are inaugurated by a growing sense, often restricted to a segment of the political community, that existing institutions have ceased adequately to meet the problems posed by an environment that they have in part created. In much the same way, scientific revolutions are inaugurated by a growing sense, again often restricted to a narrow subdivision of the scientific community, that an existing paradigm has ceased to function adequately in the exploration of an aspect of nature to which that paradigm itself had previously led the way. In both political and scientific development the sense of malfunction that can lead to crisis is prerequisite to revolution. Furthermore, though it admittedly strains the metaphor, that parallelism holds not only for the major paradigm changes, like those attributable to Copernicus and Lavoisier, but also for the far smaller ones associated with the assimilation of a new sort of phenomenon, like oxygen or X-rays. Scientific revolutions, as we noted at the end of Section V, need seem revolutionary only to those whose paradigms are affected by them. To outsiders they may, like the Balkan revolutions of the early twentieth century, seem normal parts of the developmental process. Astronomers, for example, could accept X-rays as a mere addition to knowledge, for their paradigms were unaffected by the existence of the new radiation. But for men like Kelvin, Crookes, and Roentgen, whose research dealt with radiation theory or with cathode ray tubes, the emergence of X-rays necessarily violated one paradigm as it created another. That is why these rays could be discovered only through something's first going wrong with normal research.

This genetic aspect of the parallel between political and scientific development should no longer be open to doubt. The parallel has, however, a second and more profound aspect upon which the significance of the first depends. Political revolutions aim to change political institutions in ways that those institutions themselves prohibit. Their success therefore necessitates the partial relinquishment of one set of institutions in favor of another, and in the interim, society is not fully governed by institutions at all. Initially it is crisis alone that attenuates the role of political institutions as we have already seen it attenuate the role of paradigms. In increasing numbers individuals become increasingly estranged from political life and behave more and more eccentrically within it. Then, as the crisis deepens, many of these individuals commit themselves to some concrete proposal for the reconstruction of society in a new institutional framework. At that point the society is divided into competing camps or parties, one seeking to defend the old institutional constellation, the others seeking to institute some new one. And, once that polarization has occurred, *political recourse fails.* Because they differ about the institutional matrix within which political change is to be achieved and evaluated, because they acknowledge no supra-institutional framework for the adjudication of revolutionary difference, the parties to a revolutionary conflict must finally resort to the techniques of mass persuasion, often including force. Though revolutions have had a vital role in the evolution of political institutions, that role depends upon their being partially extrapolitical or extrainstitutional events.

The remainder of this essay aims to demonstrate that the historical study of paradigm change reveals very similar characteristics in the evolution of the sciences. Like the choice between competing political institutions, that between competing paradigms proves to be a choice between incompatible modes of community life. Because it has that character, the choice is not and cannot be determined merely by the evaluative procedures characteristic of normal science, for these depend in part upon a particular paradigm, and that paradigm is at issue. When paradigms enter, as they must, into a debate about paradigm choice, their role is necessarily circular. Each group uses its own paradigm to argue in that paradigm's defense.

The resulting circularity does not, of course, make the arguments wrong or even ineffectual. The man who premises a paradigm when arguing in its defense can nonetheless provide a clear exhibit of what scientific practice will be like for those who adopt the new view of nature. That exhibit can be immensely persuasive, often compellingly so. Yet, whatever its force, the status of the circular argument is only that of persuasion. It cannot be made logically or even probabilistically compelling for those who refuse to step into the circle. The premises and values shared by the two parties to a debate over paradigms are not sufficiently extensive for that. As in political revolutions, so in paradigm choice—there is no standard higher than the assent of the relevant community. To discover how scientific revolutions are effected, we shall therefore have to examine not only the impact of nature and of logic, but also the techniques of persuasive argumentation effective within the quite special groups that constitute the community of scientists.

To discover why this issue of paradigm choice can never be unequivocally settled by logic and experiment alone, we must shortly examine the nature of the differences that separate the proponents of a traditional paradigm from their revolutionary successors. That examination is the principal object of this section and the next. We have, however, already noted numerous examples of such differences, and no one will doubt that history can supply many others. What is more likely to be doubted than their existence—and what must therefore be considered first—is that such examples provide essential information about the nature of science. Granting that paradigm rejection has been a historic fact, does it illuminate more than human credulity and confusion? Are there intrinsic reasons why the assimilation of either a new sort of phenomenon or a new scientific theory must demand the rejection of an older paradigm?

First notice that if there are such reasons, they do not derive from the logical structure of scientific knowledge. In principle, a new phenomenon might emerge without reflecting destructively upon any part of past scientific practice. Though discovering life on the moon would today be destructive of existing paradigms (these tell us things about the moon that seem incompatible with life's existence there), discovering life in some less well-known part of the galaxy would not. By the same token, a new theory does not have to conflict with any of its predecessors. It might deal exclusively with phenomena not previously known, as the quantum theory deals (but, significantly, not exclusively) with subatomic phenomena unknown before the twentieth century. Or again, the new theory might be simply a higher level theory than those known before, one that linked together a whole group of lower level theories without substantially changing any. Today, the theory of energy conservation provides just such links between dynamics, chemistry, electricity, optics, thermal theory, and so on. Still other compatible relationships between old and new theories can be conceived. Any and all of them might be exemplified by the historical process through which science has developed. If they were, scientific development would be genuinely cumulative. New sorts of phenomena would simply disclose order in an aspect of nature where none had been seen before. In the evolution of science new knowledge would replace ignorance rather than replace knowledge of another and incompatible sort.

Of course, science (or some other enterprise, perhaps less effective) might have developed in that fully cumulative manner. Many people have believed that it did so, and most still seem to suppose that cumulation is at least the ideal that historical development would display if only it had not so often been distorted by human idiosyncrasy. There are important reasons for that belief. In Section X we shall discover how closely the view of science-as-cumulation is entangled with a dominant epistemology that takes knowledge to be a construction placed directly upon raw sense data by the mind. And in Section XI we shall examine the strong support provided to the same historiographic schema by the techniques of effective science pedagogy. Nevertheless, despite the immense plausibility of that ideal image, there is increasing reason to wonder whether it can possibly be an image of *science*. After the pre-paradigm period the assimilation of all new theories and of almost all new sorts of phenomena has in fact demanded the destruction of a prior paradigm and a consequent conflict between competing schools of scientific thought. Cumulative acquisition of unanticipated novelties proves to be an almost non-existent exception to the rule of scientific development. The man who takes historic fact seriously must suspect that science does not tend toward the ideal that our image of its cumulativeness has suggested. Perhaps it is another sort of enterprise.

If, however, resistant facts can carry us that far, then a second look at the ground we have already covered may suggest that cumulative acquisition of novelty is not only rare in fact but improbable in principle. Normal research, which *is* cumulative, owes its success to the ability of scientists regularly to select problems that can be solved with conceptual and instrumental techniques close to those already in existence. (That is why an excessive concern with useful problems, regardless of their relation to existing knowledge and technique, can so easily inhibit scientific development.) The man who is striving to solve a problem defined by existing knowledge and technique is not, however, just looking around. He knows what he wants to achieve, and he designs his instruments and directs his thoughts accordingly. Unanticipated novelty, the new discovery, can emerge only to the extent that his anticipations about nature and his instruments prove wrong. Often the importance of the resulting discovery will itself be proportional to the extent and stubbornness of the anomaly that foreshadowed it. Obviously, then, there must be a conflict between the paradigm that discloses anomaly and the one that later renders the anomaly lawlike. The examples of discovery through paradigm destruction examined in Section VI did not confront us with mere historical accident. There is no other effective way in which discoveries might be generated.

The same argument applies even more clearly to the invention of new theories. There are, in principle, only three types of phenomena about which a new theory might be developed. The first consists of phenomena already well explained by existing paradigms, and these seldom provide either motive or point of departure for theory construction. When they do, as with the three famous anticipations discussed at the end of Section VII, the theories that result are seldom accepted, because nature provides no ground for discrimination. A second class of phenomena consists of those whose nature is indicated by existing paradigms but whose details can be understood only through further theory articulation. These are the phenomena to which scientists direct their research much of the time, but that research aims at the articulation of existing paradigms rather than at the invention of new ones. Only when these attempts at articulation fail do scientists encounter the third type of phenomena, the recognized anomalies whose characteristic feature is their stubborn refusal to be assimilated to existing paradigms. This type alone gives rise to new theories. Paradigms provide all phenomena except anomalies with a theory-determined place in the scientist's field of vision.

But if new theories are called forth to resolve anomalies in the relation of an existing theory to nature, then the successful new theory must somewhere permit predictions that are different from those derived from its predecessor. That difference could not occur if the two were logically compatible. In the process of being assimilated, the second must displace the first. Even a theory like energy conservation, which today seems a logical superstructure that relates to nature only through independently established theories, did not develop historically without paradigm destruction. Instead, it emerged from a crisis in which an essential ingredient was the incompatibility between Newtonian dynamics and some recently formulated consequences of the caloric theory of heat. Only after the caloric theory had been rejected could energy conservation become part of science.[1] And only after it had been part of science for some time could it come to seem a theory of a logically higher type, one not in conflict with its predecessors. It is hard to see how new theories could arise without these destructive changes in beliefs about nature. Though logical inclusiveness remains a permissible view of the relation between successive scientific theories, it is a historical implausibility.

A century ago it would, I think, have been possible to let the case for the necessity of revolutions rest at this point. But today, unfortunately, that cannot be done because the view of the subject developed above cannot be maintained if the most prevalent contemporary interpretation of the nature and function of scientific theory is accepted. That interpretation, closely associated with early logical positivism and not categorically rejected by its successors, would restrict the range and meaning of an accepted theory so that it could not possibly conflict with any later theory that made predictions about some of the same natural phenomena. The best-known and the strongest case for this restricted conception of a scientific theory emerges in discussions of the relation between contemporary Einsteinian dynamics and the older dynamical equations that descend from Newton's *Principia.* From the viewpoint of this essay these two theories are fundamentally incompatible in the sense illustrated by the relation of Copernican to Ptolemaic astronomy: Einstein's theory can be accepted only with the recognition that Newton's was wrong. Today this remains a minority view.[2] We must therefore examine the most prevalent objections to it.

[1] Silvanus P. Thompson, *Life of William Thomson Baron Kelvin of Largs* (London, 1910), I, 266–81.

[2] See, for example, the remarks by P. P. Wiener in *Philosophy of Science,* XXV (1958), 298.

The gist of these objections can be developed as follows. Relativistic dynamics cannot have shown Newtonian dynamics to be wrong, for Newtonian dynamics is still used with great success by most engineers and, in selected applications, by many physicists. Furthermore, the propriety of this use of the older theory can be proved from the very theory that has, in other applications, replaced it. Einstein's theory can be used to show that predictions from Newton's equations will be as good as our measuring instruments in all applications that satisfy a small number of restrictive conditions. For example, if Newtonian theory is to provide a good approximate solution, the relative velocities of the bodies considered must be small compared with the velocity of light. Subject to this condition and a few others, Newtonian theory seems to be derivable from Einsteinian, of which it is therefore a special case.

But, the objection continues, no theory can possibly conflict with one of its special cases. If Einsteinian science seems to make Newtonian dynamics wrong, that is only because some Newtonians were so incautious as to claim that Newtonian theory yielded entirely precise results or that it was valid at very high relative velocities. Since they could not have had any evidence for such claims, they betrayed the standards of science when they made them. In so far as Newtonian theory was ever a truly scientific theory supported by valid evidence, it still is. Only extravagant claims for the theory—claims that were never properly parts of science—can have been shown by Einstein to be wrong. Purged of these merely human extravagances, Newtonian theory has never been challenged and cannot be.

Some variant of this argument is quite sufficient to make any theory ever used by a significant group of competent scientists immune to attack. The much-maligned phlogiston theory, for example, gave order to a large number of physical and chemical phenomena. It explained why bodies burned—they were rich in phlogiston—and why metals had so many more properties in common than did their ores. The metals were all compounded from different elementary earths combined with phlogiston, and the latter, common to all metals, produced common properties. In addition, the phlogiston theory accounted for a number of reactions in which acids were formed by the combustion of substances like carbon and sulphur. Also, it explained the decrease of volume when combustion occurs in a confined volume of air—the phlogiston released by combustion "spoils" the elasticity of the air that absorbed it, just as fire "spoils" the elasticity of a steel spring.[3] If these were the only phenomena that the phlogiston theorists had claimed for their theory, that theory could never have been challenged. A similar argument will suffice for any theory that has ever been successfully applied to any range of phenomena at all.

But to save theories in this way, their range of application must be restricted to those phenomena and to that precision of observation with which the experimental evidence in hand already deals.[4] Carried just a step further (and the step can scarcely be avoided once the first is taken), such a limitation prohibits the scientist from claiming to speak "scientifically" about any phenomenon not already observed. Even in its present form the restriction forbids the scientist to rely upon a theory in his own research whenever that research enters an area or seeks a degree of precision for which past practice with the theory offers no precedent. These prohibitions are logically unexceptionable. But the result of accepting them would be the end of the research through which science may develop further.

By now that point too is virtually a tautology. Without commitment to a paradigm there could be no normal science. Furthermore, that commitment must extend to areas and to degrees of precision for which there is no full precedent. If it did not, the paradigm could provide no puzzles that had not already been solved. Besides, it is not only normal science that depends upon commitment to a paradigm. If existing theory binds the

[3] James B. Conant, *Overthrow of the Phlogiston Theory* (Cambridge, 1950), pp. 13–16; and J. R. Partington, *A Short History of Chemistry* (2d ed.; London, 1951), pp. 85–88. The fullest and most sympathetic account of the phlogiston theory's achievements is by H. Metzger, *Newton, Stahl, Boerhaave et la doctrine chimique* (Paris, 1930), Part II.

[4] Compare the conclusions reached through a very different sort of analysis by R. B. Braithwaite, *Scientific Explanation* (Cambridge, 1953), pp. 50–87, esp. p. 76.

scientist only with respect to existing applications, then there can be no surprises, anomalies, or crises. But these are just the signposts that point the way to extraordinary science. If positivistic restrictions on the range of a theory's legitimate applicability are taken literally, the mechanism that tells the scientific community what problems may lead to fundamental change must cease to function. And when that occurs, the community will inevitably return to something much like its pre-paradigm state, a condition in which all members practice science but in which their gross product scarcely resembles science at all. Is it really any wonder that the price of significant scientific advance is a commitment that runs the risk of being wrong?

More important, there is a revealing logical lacuna in the positivist's argument, one that will reintroduce us immediately to the nature of revolutionary change. Can Newtonian dynamics really be *derived* from relativistic dynamics? What would such a derivation look like? Imagine a set of statements, $E_1, E_2, \ldots, E_n$, which together embody the laws of relativity theory. These statements contain variables and parameters representing spatial position, time, rest mass, etc. From them, together with the apparatus of logic and mathematics, is deducible a whole set of further statements including some that can be checked by observation. To prove the adequacy of Newtonian dynamics as a special case, we must add to the E_i's additional statements, like $(v/c)^2 << 1$, restricting the range of the parameters and variables. This enlarged set of statements is then manipulated to yield a new set, $N_1, N_2, \ldots, N_m$, which is identical in form with Newton's laws of motion, the law of gravity, and so on. Apparently Newtonian dynamics has been derived from Einsteinian, subject to a few limiting conditions.

Yet the derivation is spurious, at least to this point. Though the N_i's are a special case of the laws of relativistic mechanics, they are not Newton's Laws. Or at least they are not unless those laws are reinterpreted in a way that would have been impossible until after Einstein's work. The variables and parameters that in the Einsteinian E_i's represented spatial position, time, mass, etc., still occur in the N_i's; and they there still represent Einsteinian space, time, and mass. But the physical referents of these Einsteinian concepts are by no means identical with those of the Newtonian concepts that bear the same name. (Newtonian mass is conserved; Einsteinian is convertible with energy. Only at low relative velocities may the two be measured in the same way, and even then they must not be conceived to be the same.) Unless we change the definitions of the variables in the N_i's, the statements we have derived are not Newtonian. If we do change them, we cannot properly be said to have *derived* Newton's Laws, at least not in any sense of "derive" now generally recognized. Our argument has, of course, explained why Newton's Laws ever seemed to work. In doing so it has justified, say, an automobile driver in acting as though he lived in a Newtonian universe. An argument of the same type is used to justify teaching earth-centered astronomy to surveyors. But the argument has still not done what it purported to do. It has not, that is, shown Newton's Laws to be a limiting case of Einstein's. For in the passage to the limit it is not only the forms of the laws that have changed. Simultaneously we have had to alter the fundamental structural elements of which the universe to which they apply is composed.

This need to change the meaning of established and familiar concepts is central to the revolutionary impact of Einstein's theory. Though subtler than the changes from geocentrism to heliocentrism, from phlogiston to oxygen, or from corpuscles to waves, the resulting conceptual transformation is no less decisively destructive of a previously established paradigm. We may even come to see it as a prototype for revolutionary reorientations in the sciences. Just because it did not involve the introduction of additional objects or concepts, the transition from Newtonian to Einsteinian mechanics illustrates with particular clarity the scientific revolution as a displacement of the conceptual network through which scientists view the world.

These remarks should suffice to show what might, in another philosophical climate, have been taken for granted. At least for scientists, most of the apparent differences between a discarded scientific theory and its successor are real. Though an out-of-date theory can always be viewed as a special case of its up-to-date successor, it must be transformed for the purpose. And the transformation is one that can be undertaken only with the advantages of hindsight, the explicit guidance of the more recent theory. Furthermore, even if that transformation were a legitimate device to employ in interpreting the older theory, the result of its application would be a theory so restricted that it could only restate what was already known. Because of its economy, that restatement would have utility, but it could not suffice for the guidance of research.

Let us, therefore, now take it for granted that the differences between successive paradigms are both necessary and irreconcilable. Can we then say more explicitly what sorts of differences these are? The most apparent type has already been illustrated repeatedly. Successive paradigms tell us different things about the population of the universe and about that population's behavior. They differ, that is, about such questions as the existence of subatomic particles, the materiality of light, and the conservation of heat or of energy. These are the substantive differences between successive paradigms, and they require no further illustration. But paradigms differ in more than substance, for they are directed not only to nature but also back upon the science that produced them. They are the source of the methods, problem-field, and standards of solution accepted by any mature scientific community at any given time. As a result, the reception of a new paradigm often necessitates a redefinition of the corresponding science. Some old problems may be relegated to another science or declared entirely "unscientific." Others that were previously non-existent or trivial may, with a new paradigm, become the very archetypes of significant scientific achievement. And as the problems change, so, often, does the standard that distinguishes a real scientific solution from a mere metaphysical speculation, word game, or mathematical play. The normal-scientific tradition that emerges from a scientific revolution is not only incompatible but often actually incommensurable with that which has gone before.

The impact of Newton's work upon the normal seventeenth-century tradition of scientific practice provides a striking example of these subtler effects of paradigm shift. Before Newton was born the "new science" of the century had at last succeeded in rejecting Aristotelian and scholastic explanations expressed in terms of the essences of material bodies. To say that a stone fell because its "nature" drove it toward the center of the universe had been made to look a mere tautological word-play, something it had not previously been. Henceforth the entire flux of sensory appearances, including color, taste, and even weight, was to be explained in terms of the size, shape, position, and motion of the elementary corpuscles of base matter. The attribution of other qualities to the elementary atoms was a resort to the occult and therefore out of bounds for science. Molière caught the new spirit precisely when he ridiculed the doctor who explained opium's efficacy as a soporific by attributing to it a dormitive potency. During the last half of the seventeenth century many scientists preferred to say that the round shape of the opium particles enabled them to sooth the nerves about which they moved.[5]

In an earlier period explanations in terms of occult qualities had been an integral part of productive scientific work. Nevertheless, the seventeenth century's new commitment to mechanico-corpuscular explanation proved immensely fruitful for a number of sciences, ridding them of problems that had defied generally accepted solution and suggesting others to replace them. In dynamics, for example, Newton's three laws of motion are less a product of novel experiments than of the attempt to reinterpret well-known observations in terms of the motions and interactions of primary neutral corpuscles. Consider just one concrete illustration. Since neutral corpuscles could act on each other only by contact, the mechanico-corpuscular view of nature directed scientific attention to a brand-new subject of study, the alteration of particulate motions by collisions. Descartes announced the problem and provided its first putative

[5] For corpuscularism in general, see Marie Boas, "The Establishment of the Mechanical Philosophy," *Osiris*, X (1952), 412–541. For the effect of particle-shape on taste, see *ibid.*, p. 483.

solution. Huyghens, Wren, and Wallis carried it still further, partly by experimenting with colliding pendulum bobs, but mostly by applying previously well-known characteristics of motion to the new problem. And Newton embedded their results in his laws of motion. The equal "action" and "reaction" of the third law are the changes in quantity of motion experienced by the two parties to a collision. The same change of motion supplies the definition of dynamical force implicit in the second law. In this case, as in many others during the seventeenth century, the corpuscular paradigm bred both a new problem and a large part of that problem's solution.[6]

Yet, though much of Newton's work was directed to problems and embodied standards derived from the mechanico-corpuscular world view, the effect of the paradigm that resulted from his work was a further and partially destructive change in the problems and standards legitimate for science. Gravity, interpreted as an innate attraction between every pair of particles of matter, was an occult quality in the same sense as the scholastics' "tendency to fall" had been. Therefore, while the standards of corpuscularism remained in effect, the search for a mechanical explanation of gravity was one of the most challenging problems for those who accepted the *Principia* as paradigm. Newton devoted much attention to it and so did many of his eighteenth-century successors. The only apparent option was to reject Newton's theory for its failure to explain gravity, and that alternative, too, was widely adopted. Yet neither of these views ultimately triumphed. Unable either to practice science without the *Principia* or to make that work conform to the corpuscular standards of the seventeenth century, scientists gradually accepted the view that gravity was indeed innate. By the mid-eighteenth century that interpretation had been almost universally accepted, and the result was a genuine reversion (which is not the same as a retrogression) to a scholastic standard. Innate attractions and repulsions joined size, shape, posi-

[6] R. Dugas, *La mécanique au XVIIe siècle* (Neuchatel, 1954), pp. 177–85, 284–98, 345–56.

tion, and motion as physically irreducible primary properties of matter.[7]

The resulting change in the standards and problem-field of physical science was once again consequential. By the 1740's, for example, electricians could speak of the attractive "virtue" of the electric fluid without thereby inviting the ridicule that had greeted Molière's doctor a century before. As they did so, electrical phenomena increasingly displayed an order different from the one they had shown when viewed as the effects of a mechanical effluvium that could act only by contact. In particular, when electrical action-at-a-distance became a subject for study in its own right, the phenomenon we now call charging by induction could be recognized as one of its effects. Previously, when seen at all, it had been attributed to the direct action of electrical "atmospheres" or to the leakages inevitable in any electrical laboratory. The new view of inductive effects was, in turn, the key to Franklin's analysis of the Leyden jar and thus to the emergence of a new and Newtonian paradigm for electricity. Nor were dynamics and electricity the only scientific fields affected by the legitimization of the search for forces innate to matter. The large body of eighteenth-century literature on chemical affinities and replacement series also derives from this supramechanical aspect of Newtonianism. Chemists who believed in these differential attractions between the various chemical species set up previously unimagined experiments and searched for new sorts of reactions. Without the data and the chemical concepts developed in that process, the later work of Lavoisier and, more particularly, of Dalton would be incomprehensible.[8] Changes in the standards governing permissible problems, concepts, and explanations can transform a science. In the next section I shall even suggest a sense in which they transform the world.

[7] I. B. Cohen, *Franklin and Newton: An Inquiry into Speculative Newtonian Experimental Science and Franklin's Work in Electricity as an Example Thereof* (Philadelphia, 1956), chaps. vi–vii.

[8] For electricity, see *ibid*, chaps. viii–ix. For chemistry, see Metzger, *op. cit.*, Part I.

Other examples of these nonsubstantive differences between successive paradigms can be retrieved from the history of any science in almost any period of its development. For the moment let us be content with just two other and far briefer illustrations. Before the chemical revolution, one of the acknowledged tasks of chemistry was to account for the qualities of chemical substances and for the changes these qualities underwent during chemical reactions. With the aid of a small number of elementary "principles"—of which phlogiston was one—the chemist was to explain why some substances are acidic, others metalline, combustible, and so forth. Some success in this direction had been achieved. We have already noted that phlogiston explained why the metals were so much alike, and we could have developed a similar argument for the acids. Lavoisier's reform, however, ultimately did away with chemical "principles," and thus ended by depriving chemistry of some actual and much potential explanatory power. To compensate for this loss, a change in standards was required. During much of the nineteenth century failure to explain the qualities of compounds was no indictment of a chemical theory.[9]

Or again, Clerk Maxwell shared with other nineteenth-century proponents of the wave theory of light the conviction that light waves must be propagated through a material ether. Designing a mechanical medium to support such waves was a standard problem for many of his ablest contemporaries. His own theory, however, the electromagnetic theory of light, gave no account at all of a medium able to support light waves, and it clearly made such an account harder to provide than it had seemed before. Initially, Maxwell's theory was widely rejected for those reasons. But, like Newton's theory, Maxwell's proved difficult to dispense with, and as it achieved the status of a paradigm, the community's attitude toward it changed. In the early decades of the twentieth century Maxwell's insistence upon the existence of a mechanical ether looked more and more like lip service, which it emphatically had not been, and the attempts to design such an ethereal medium were abandoned. Scientists no

[9] E. Meyerson, *Identity and Reality* (New York, 1930), chap. x.

longer thought it unscientific to speak of an electrical "displacement" without specifying what was being displaced. The result, again, was a new set of problems and standards, one which, in the event, had much to do with the emergence of relativity theory.[10]

These characteristic shifts in the scientific community's conception of its legitimate problems and standards would have less significance to this essay's thesis if one could suppose that they always occurred from some methodologically lower to some higher type. In that case their effects, too, would seem cumulative. No wonder that some historians have argued that the history of science records a continuing increase in the maturity and refinement of man's conception of the nature of science.[11] Yet the case for cumulative development of science's problems and standards is even harder to make than the case for cumulation of theories. The attempt to explain gravity, though fruitfully abandoned by most eighteenth-century scientists, was not directed to an intrinsically illegitimate problem; the objections to innate forces were neither inherently unscientific nor metaphysical in some pejorative sense. There are no external standards to permit a judgment of that sort. What occurred was neither a decline nor a raising of standards, but simply a change demanded by the adoption of a new paradigm. Furthermore, that change has since been reversed and could be again. In the twentieth century Einstein succeeded in explaining gravitational attractions, and that explanation has returned science to a set of canons and problems that are, in this particular respect, more like those of Newton's predecessors than of his successors. Or again, the development of quantum mechanics has reversed the methodological prohibition that originated in the chemical revolution. Chemists now attempt, and with great success, to explain the color, state of aggregation, and other qualities of the substances used and produced in their laboratories. A similar rever-

[10] E. T. Whittaker, *A History of the Theories of Aether and Electricity*, II (London, 1953), 28–30.

[11] For a brilliant and entirely up-to-date attempt to fit scientific development into this Procrustean bed, see C. C. Gillispie, *The Edge of Objectivity: An Essay in the History of Scientific Ideas* (Princeton, 1960).

sal may even be underway in electromagnetic theory. Space, in contemporary physics, is not the inert and homogenous substratum employed in both Newton's and Maxwell's theories; some of its new properties are not unlike those once attributed to the ether; we may someday come to know what an electric displacement is.

By shifting emphasis from the cognitive to the normative functions of paradigms, the preceding examples enlarge our understanding of the ways in which paradigms give form to the scientific life. Previously, we had principally examined the paradigm's role as a vehicle for scientific theory. In that role it functions by telling the scientist about the entities that nature does and does not contain and about the ways in which those entities behave. That information provides a map whose details are elucidated by mature scientific research. And since nature is too complex and varied to be explored at random, that map is as essential as observation and experiment to science's continuing development. Through the theories they embody, paradigms prove to be constitutive of the research activity. They are also, however, constitutive of science in other respects, and that is now the point. In particular, our most recent examples show that paradigms provide scientists not only with a map but also with some of the directions essential for map-making. In learning a paradigm the scientist acquires theory, methods, and standards together, usually in an inextricable mixture. Therefore, when paradigms change, there are usually significant shifts in the criteria determining the legitimacy both of problems and of proposed solutions.

That observation returns us to the point from which this section began, for it provides our first explicit indication of why the choice between competing paradigms regularly raises questions that cannot be resolved by the criteria of normal science. To the extent, as significant as it is incomplete, that two scientific schools disagree about what is a problem and what a solution, they will inevitably talk through each other when debating the relative merits of their respective paradigms. In the partially circular arguments that regularly result, each paradigm will be shown to satisfy more or less the criteria that it dictates for itself and to fall short of a few of those dictated by its opponent. There are other reasons, too, for the incompleteness of logical contact that consistently characterizes paradigm debates. For example, since no paradigm ever solves all the problems it defines and since no two paradigms leave all the same problems unsolved, paradigm debates always involve the question: Which problems is it more significant to have solved? Like the issue of competing standards, that question of values can be answered only in terms of criteria that lie outside of normal science altogether, and it is that recourse to external criteria that most obviously makes paradigm debates revolutionary. Something even more fundamental than standards and values is, however, also at stake. I have so far argued only that paradigms are constitutive of science. Now I wish to display a sense in which they are constitutive of nature as well.

X. Revolutions as Changes of World View

Examining the record of past research from the vantage of contemporary historiography, the historian of science may be tempted to exclaim that when paradigms change, the world itself changes with them. Led by a new paradigm, scientists adopt new instruments and look in new places. Even more important, during revolutions scientists see new and different things when looking with familiar instruments in places they have looked before. It is rather as if the professional community had been suddenly transported to another planet where familiar objects are seen in a different light and are joined by unfamiliar ones as well. Of course, nothing of quite that sort does occur: there is no geographical transplantation; outside the laboratory everyday affairs usually continue as before. Nevertheless, paradigm changes do cause scientists to see the world of their research-engagement differently. In so far as their only recourse to that world is through what they see and do, we may want to say that after a revolution scientists are responding to a different world.

It is as elementary prototypes for these transformations of the scientist's world that the familiar demonstrations of a switch in visual gestalt prove so suggestive. What were ducks in the scientist's world before the revolution are rabbits afterwards. The man who first saw the exterior of the box from above later sees its interior from below. Transformations like these, though usually more gradual and almost always irreversible, are common concomitants of scientific training. Looking at a contour map, the student sees lines on paper, the cartographer a picture of a terrain. Looking at a bubble-chamber photograph, the student sees confused and broken lines, the physicist a record of familiar subnuclear events. Only after a number of such transformations of vision does the student become an inhabitant of the scientist's world, seeing what the scientist sees and responding as the scientist does. The world that the student then enters is not, however, fixed once and for all by the nature of the environment, on the one hand, and of science, on the other. Rather, it is determined jointly by the environment and the particular normal-scientific tradition that the student has been trained to pursue. Therefore, at times of revolution, when the normal-scientific tradition changes, the scientist's perception of his environment must be re-educated—in some familiar situations he must learn to see a new gestalt. After he has done so the world of his research will seem, here and there, incommensurable with the one he had inhabited before. That is another reason why schools guided by different paradigms are always slightly at cross-purposes.

In their most usual form, of course, gestalt experiments illustrate only the nature of perceptual transformations. They tell us nothing about the role of paradigms or of previously assimilated experience in the process of perception. But on that point there is a rich body of psychological literature, much of it stemming from the pioneering work of the Hanover Institute. An experimental subject who puts on goggles fitted with inverting lenses initially sees the entire world upside down. At the start his perceptual apparatus functions as it had been trained to function in the absence of the goggles, and the result is extreme disorientation, an acute personal crisis. But after the subject has begun to learn to deal with his new world, his entire visual field flips over, usually after an intervening period in which vision is simply confused. Thereafter, objects are again seen as they had been before the goggles were put on. The assimilation of a previously anomalous visual field has reacted upon and changed the field itself.[1] Literally as well as metaphorically, the man accustomed to inverting lenses has undergone a revolutionary transformation of vision.

The subjects of the anomalous playing-card experiment discussed in Section VI experienced a quite similar transformation. Until taught by prolonged exposure that the universe contained

[1] The original experiments were by George M. Stratton, "Vision without Inversion of the Retinal Image," *Psychological Review,* IV (1897), 341–60, 463–81. A more up-to-date review is provided by Harvey A. Carr, *An Introduction to Space Perception* (New York, 1935), pp. 18–57.

anomalous cards, they saw only the types of cards for which previous experience had equipped them. Yet once experience had provided the requisite additional categories, they were able to see all anomalous cards on the first inspection long enough to permit any identification at all. Still other experiments demonstrate that the perceived size, color, and so on, of experimentally displayed objects also varies with the subject's previous training and experience.[2] Surveying the rich experimental literature from which these examples are drawn makes one suspect that something like a paradigm is prerequisite to perception itself. What a man sees depends both upon what he looks at and also upon what his previous visual-conceptual experience has taught him to see. In the absence of such training there can only be, in William James's phrase, "a bloomin' buzzin' confusion."

In recent years several of those concerned with the history of science have found the sorts of experiments described above immensely suggestive. N. R. Hanson, in particular, has used gestalt demonstrations to elaborate some of the same consequences of scientific belief that concern me here.[3] Other colleagues have repeatedly noted that history of science would make better and more coherent sense if one could suppose that scientists occasionally experienced shifts of perception like those described above. Yet, though psychological experiments are suggestive, they cannot, in the nature of the case, be more than that. They do display characteristics of perception that *could* be central to scientific development, but they do not demonstrate that the careful and controlled observation exercised by the research scientist at all partakes of those characteristics. Furthermore, the very nature of these experiments makes any direct demonstration of that point impossible. If historical example is to make these psychological experiments seem rele-

[2] For examples, see Albert H. Hastorf, "The Influence of Suggestion on the Relationship between Stimulus Size and Perceived Distance," *Journal of Psychology,* XXIX (1950), 195–217; and Jerome S. Bruner, Leo Postman, and John Rodrigues, "Expectations and the Perception of Color," *American Journal of Psychology,* LXIV (1951), 216–27.

[3] N. R. Hanson, *Patterns of Discovery* (Cambridge, 1958), chap. i.

vant, we must first notice the sorts of evidence that we may and may not expect history to provide.

The subject of a gestalt demonstration knows that his perception has shifted because he can make it shift back and forth repeatedly while he holds the same book or piece of paper in his hands. Aware that nothing in his environment has changed, he directs his attention increasingly not to the figure (duck or rabbit) but to the lines on the paper he is looking at. Ultimately he may even learn to see those lines without seeing either of the figures, and he may then say (what he could not legitimately have said earlier) that it is these lines that he really sees but that he sees them alternately *as* a duck and *as* a rabbit. By the same token, the subject of the anomalous card experiment knows (or, more accurately, can be persuaded) that his perception must have shifted because an external authority, the experimenter, assures him that regardless of what he *saw,* he was *looking at* a black five of hearts all the time. In both these cases, as in all similar psychological experiments, the effectiveness of the demonstration depends upon its being analyzable in this way. Unless there were an external standard with respect to which a switch of vision could be demonstrated, no conclusion about alternate perceptual possibilities could be drawn.

With scientific observation, however, the situation is exactly reversed. The scientist can have no recourse above or beyond what he sees with his eyes and instruments. If there were some higher authority by recourse to which his vision might be shown to have shifted, then that authority would itself become the source of his data, and the behavior of his vision would become a source of problems (as that of the experimental subject is for the psychologist). The same sorts of problems would arise if the scientist could switch back and forth like the subject of the gestalt experiments. The period during which light was "sometimes a wave and sometimes a particle" was a period of crisis—a period when something was wrong—and it ended only with the development of wave mechanics and the realization that light was a self-consistent entity different from both waves and particles. In the sciences, therefore, if perceptual switches accompany paradigm changes, we may not expect scientists to attest to these changes directly. Looking at the moon, the convert to Copernicanism does not say, "I used to see a planet, but now I see a satellite." That locution would imply a sense in which the Ptolemaic system had once been correct. Instead, a convert to the new astronomy says, "I once took the moon to be (or saw the moon as) a planet, but I was mistaken." That sort of statement does recur in the aftermath of scientific revolutions. If it ordinarily disguises a shift of scientific vision or some other mental transformation with the same effect, we may not expect direct testimony about that shift. Rather we must look for indirect and behavioral evidence that the scientist with a new paradigm sees differently from the way he had seen before.

Let us then return to the data and ask what sorts of transformations in the scientist's world the historian who believes in such changes can discover. Sir William Herschel's discovery of Uranus provides a first example and one that closely parallels the anomalous card experiment. On at least seventeen different occasions between 1690 and 1781, a number of astronomers, including several of Europe's most eminent observers, had seen a star in positions that we now suppose must have been occupied at the time by Uranus. One of the best observers in this group had actually seen the star on four successive nights in 1769 without noting the motion that could have suggested another identification. Herschel, when he first observed the same object twelve years later, did so with a much improved telescope of his own manufacture. As a result, he was able to notice an apparent disk-size that was at least unusual for stars. Something was awry, and he therefore postponed identification pending further scrutiny. That scrutiny disclosed Uranus' motion among the stars, and Herschel therefore announced that he had seen a new comet! Only several months later, after fruitless attempts to fit the observed motion to a cometary orbit, did Lexell suggest that the orbit was probably planetary.[4] When that suggestion was accepted, there were several fewer stars and one more planet in the world of the professional astronomer. A celestial body that

[4] Peter Doig, *A Concise History of Astronomy* (London, 1950), pp. 115–16.

had been observed off and on for almost a century was seen differently after 1781 because, like an anomalous playing card, it could no longer be fitted to the perceptual categories (star or comet) provided by the paradigm that had previously prevailed.

The shift of vision that enabled astronomers to see Uranus, the planet, does not, however, seem to have affected only the perception of that previously observed object. Its consequences were more far-reaching. Probably, though the evidence is equivocal, the minor paradigm change forced by Herschel helped to prepare astronomers for the rapid discovery, after 1801, of the numerous minor planets or asteroids. Because of their small size, these did not display the anomalous magnification that had alerted Herschel. Nevertheless, astronomers prepared to find additional planets were able, with standard instruments, to identify twenty of them in the first fifty years of the nineteenth century.[5] The history of astronomy provides many other examples of paradigm-induced changes in scientific perception, some of them even less equivocal. Can it conceivably be an accident, for example, that Western astronomers first saw change in the previously immutable heavens during the half-century after Copernicus' new paradigm was first proposed? The Chinese, whose cosmological beliefs did not preclude celestial change, had recorded the appearance of many new stars in the heavens at a much earlier date. Also, even without the aid of a telescope, the Chinese had systematically recorded the appearance of sunspots centuries before these were seen by Galileo and his contemporaries.[6] Nor were sunspots and a new star the only examples of celestial change to emerge in the heavens of Western astronomy immediately after Copernicus. Using traditional instruments, some as simple as a piece of thread, late sixteenth-century astronomers repeatedly discovered that comets wandered at will through the space previously reserved for the

[5] Rudolph Wolf, *Geschichte der Astronomie* (Munich, 1877), pp. 513–15, 683–93. Notice particularly how difficult Wolf's account makes it to explain these discoveries as a consequence of Bode's Law.

[6] Joseph Needham, *Science and Civilization in China*, III (Cambridge, 1959), 423–29, 434–36.

immutable planets and stars.[7] The very ease and rapidity with which astronomers saw new things when looking at old objects with old instruments may make us wish to say that, after Copernicus, astronomers lived in a different world. In any case, their research responded as though that were the case.

The preceding examples are selected from astronomy because reports of celestial observation are frequently delivered in a vocabulary consisting of relatively pure observation terms. Only in such reports can we hope to find anything like a full parallelism between the observations of scientists and those of the psychologist's experimental subjects. But we need not insist on so full a parallelism, and we have much to gain by relaxing our standard. If we can be content with the everyday use of the verb 'to see,' we may quickly recognize that we have already encountered many other examples of the shifts in scientific perception that accompany paradigm change. The extended use of 'perception' and of 'seeing' will shortly require explicit defense, but let me first illustrate its application in practice.

Look again for a moment at two of our previous examples from the history of electricity. During the seventeenth century, when their research was guided by one or another effluvium theory, electricians repeatedly saw chaff particles rebound from, or fall off, the electrified bodies that had attracted them. At least that is what seventeenth-century observers said they saw, and we have no more reason to doubt their reports of perception than our own. Placed before the same apparatus, a modern observer would see electrostatic repulsion (rather than mechanical or gravitational rebounding), but historically, with one universally ignored exception, electrostatic repulsion was not seen as such until Hauksbee's large-scale apparatus had greatly magnified its effects. Repulsion after contact electrification was, however, only one of many new repulsive effects that Hauksbee saw. Through his researches, rather as in a gestalt switch, repulsion suddenly became *the* fundamental manifestation of electrification, and it was then attraction that needed to be ex-

[7] T. S. Kuhn, *The Copernican Revolution* (Cambridge, Mass., 1957), pp. 206–9.

plained.[8] The electrical phenomena visible in the early eighteenth century were both subtler and more varied than those seen by observers in the seventeenth century. Or again, after the assimilation of Franklin's paradigm, the electrician looking at a Leyden jar saw something different from what he had seen before. The device had become a condenser, for which neither the jar shape nor glass was required. Instead, the two conducting coatings—one of which had been no part of the original device—emerged to prominence. As both written discussions and pictorial representations gradually attest, two metal plates with a non-conductor between them had become the prototype for the class.[9] Simultaneously, other inductive effects received new descriptions, and still others were noted for the first time.

Shifts of this sort are not restricted to astronomy and electricity. We have already remarked some of the similar transformations of vision that can be drawn from the history of chemistry. Lavoisier, we said, saw oxygen where Priestley had seen dephlogisticated air and where others had seen nothing at all. In learning to see oxygen, however, Lavoisier also had to change his view of many other more familiar substances. He had, for example, to see a compound ore where Priestley and his contemporaries had seen an elementary earth, and there were other such changes besides. At the very least, as a result of discovering oxygen, Lavoisier saw nature differently. And in the absence of some recourse to that hypothetical fixed nature that he "saw differently," the principle of economy will urge us to say that after discovering oxygen Lavoisier worked in a different world.

I shall inquire in a moment about the possibility of avoiding this strange locution, but first we require an additional example of its use, this one deriving from one of the best known parts of the work of Galileo. Since remote antiquity most people have seen one or another heavy body swinging back and forth on a string or chain until it finally comes to rest. To the Aristotelians,

[8] Duane Roller and Duane H. D. Roller, *The Development of the Concept of Electric Charge* (Cambridge, Mass., 1954), pp. 21–29.

[9] See the discussion in Section VII and the literature to which the reference there cited in note 9 will lead.

who believed that a heavy body is moved by its own nature from a higher position to a state of natural rest at a lower one, the swinging body was simply falling with difficulty. Constrained by the chain, it could achieve rest at its low point only after a tortuous motion and a considerable time. Galileo, on the other hand, looking at the swinging body, saw a pendulum, a body that almost succeeded in repeating the same motion over and over again ad infinitum. And having seen that much, Galileo observed other properties of the pendulum as well and constructed many of the most significant and original parts of his new dynamics around them. From the properties of the pendulum, for example, Galileo derived his only full and sound arguments for the independence of weight and rate of fall, as well as for the relationship between vertical height and terminal velocity of motions down inclined planes.[10] All these natural phenomena he saw differently from the way they had been seen before.

Why did that shift of vision occur? Through Galileo's individual genius, of course. But note that genius does not here manifest itself in more accurate or objective observation of the swinging body. Descriptively, the Aristotelian perception is just as accurate. When Galileo reported that the pendulum's period was independent of amplitude for amplitudes as great as 90°, his view of the pendulum led him to see far more regularity than we can now discover there.[11] Rather, what seems to have been involved was the exploitation by genius of perceptual possibilities made available by a medieval paradigm shift. Galileo was not raised completely as an Aristotelian. On the contrary, he was trained to analyze motions in terms of the impetus theory, a late medieval paradigm which held that the continuing motion of a heavy body is due to an internal power implanted in it by the projector that initiated its motion. Jean Buridan and Nicole Oresme, the fourteenth-century scholastics who brought the impetus theory to its most perfect formulations, are the first men

[10] Galileo Galilei, *Dialogues concerning Two New Sciences*, trans. H. Crew and A. de Salvio (Evanston, Ill., 1946), pp. 80–81, 162–66.

[11] *Ibid.*, pp. 91–94, 244.

known to have seen in oscillatory motions any part of what Galileo saw there. Buridan describes the motion of a vibrating string as one in which impetus is first implanted when the string is struck; the impetus is next consumed in displacing the string against the resistance of its tension; tension then carries the string back, implanting increasing impetus until the mid-point of motion is reached; after that the impetus displaces the string in the opposite direction, again against the string's tension, and so on in a symmetric process that may continue indefinitely. Later in the century Oresme sketched a similar analysis of the swinging stone in what now appears as the first discussion of a pendulum.[12] His view is clearly very close to the one with which Galileo first approached the pendulum. At least in Oresme's case, and almost certainly in Galileo's as well, it was a view made possible by the transition from the original Aristotelian to the scholastic impetus paradigm for motion. Until that scholastic paradigm was invented, there were no pendulums, but only swinging stones, for the scientist to see. Pendulums were brought into existence by something very like a paradigm-induced gestalt switch.

Do we, however, really need to describe what separates Galileo from Aristotle, or Lavoisier from Priestley, as a transformation of vision? Did these men really *see* different things when *looking at* the same sorts of objects? Is there any legitimate sense in which we can say that they pursued their research in different worlds? Those questions can no longer be postponed, for there is obviously another and far more usual way to describe all of the historical examples outlined above. Many readers will surely want to say that what changes with a paradigm is only the scientist's interpretation of observations that themselves are fixed once and for all by the nature of the environment and of the perceptual apparatus. On this view, Priestley and Lavoisier both saw oxygen, but they interpreted their observations differently; Aristotle and Galileo both saw pendu-

[12] M. Clagett, *The Science of Mechanics in the Middle Ages* (Madison, Wis., 1959), pp. 537–38, 570.

lums, but they differed in their interpretations of what they both had seen.

Let me say at once that this very usual view of what occurs when scientists change their minds about fundamental matters can be neither all wrong nor a mere mistake. Rather it is an essential part of a philosophical paradigm initiated by Descartes and developed at the same time as Newtonian dynamics. That paradigm has served both science and philosophy well. Its exploitation, like that of dynamics itself, has been fruitful of a fundamental understanding that perhaps could not have been achieved in another way. But as the example of Newtonian dynamics also indicates, even the most striking past success provides no guarantee that crisis can be indefinitely postponed. Today research in parts of philosophy, psychology, linguistics, and even art history, all converge to suggest that the traditional paradigm is somehow askew. That failure to fit is also made increasingly apparent by the historical study of science to which most of our attention is necessarily directed here.

None of these crisis-promoting subjects has yet produced a viable alternate to the traditional epistemological paradigm, but they do begin to suggest what some of that paradigm's characteristics will be. I am, for example, acutely aware of the difficulties created by saying that when Aristotle and Galileo looked at swinging stones, the first saw constrained fall, the second a pendulum. The same difficulties are presented in an even more fundamental form by the opening sentences of this section: though the world does not change with a change of paradigm, the scientist afterward works in a different world. Nevertheless, I am convinced that we must learn to make sense of statements that at least resemble these. What occurs during a scientific revolution is not fully reducible to a reinterpretation of individual and stable data. In the first place, the data are not unequivocally stable. A pendulum is not a falling stone, nor is oxygen dephlogisticated air. Consequently, the data that scientists collect from these diverse objects are, as we shall shortly see, themselves different. More important, the process by which either the individual or the community makes the transition from constrained fall to the pendulum or from dephlogisticated air to oxygen is not one that resembles interpretation. How could it do so in the absence of fixed data for the scientist to interpret? Rather than being an interpreter, the scientist who embraces a new paradigm is like the man wearing inverting lenses. Confronting the same constellation of objects as before and knowing that he does so, he nevertheless finds them transformed through and through in many of their details.

None of these remarks is intended to indicate that scientists do not characteristically interpret observations and data. On the contrary, Galileo interpreted observations on the pendulum, Aristotle observations on falling stones, Musschenbroek observations on a charge-filled bottle, and Franklin observations on a condenser. But each of these interpretations presupposed a paradigm. They were parts of normal science, an enterprise that, as we have already seen, aims to refine, extend, and articulate a paradigm that is already in existence. Section III provided many examples in which interpretation played a central role. Those examples typify the overwhelming majority of research. In each of them the scientist, by virtue of an accepted paradigm, knew what a datum was, what instruments might be used to retrieve it, and what concepts were relevant to its interpretation. Given a paradigm, interpretation of data is central to the enterprise that explores it.

But that interpretive enterprise—and this was the burden of the paragraph before last—can only articulate a paradigm, not correct it. Paradigms are not corrigible by normal science at all. Instead, as we have already seen, normal science ultimately leads only to the recognition of anomalies and to crises. And these are terminated, not by deliberation and interpretation, but by a relatively sudden and unstructured event like the gesalt switch. Scientists then often speak of the "scales falling from the eyes" or of the "lightning flash" that "inundates" a previously obscure puzzle, enabling its components to be seen in a new way that for the first time permits its solution. On other occasions the relevant illumination comes in sleep.[13] No ordinary sense of the term 'interpretation' fits these flashes of intuition through which a new paradigm is born. Though such intuitions depend upon the experience, both anomalous and congruent, gained with the old paradigm, they are not logically or piecemeal linked to particular items of that experience as an interpretation would be. Instead, they gather up large portions of that experience and transform them to the rather different bundle of experience that will thereafter be linked piecemeal to the new paradigm but not to the old.

To learn more about what these differences in experience can be, return for a moment to Aristotle, Galileo, and the pendulum. What data did the interaction of their different paradigms and their common environment make accessible to each of them? Seeing constrained fall, the Aristotelian would measure (or at least discuss—the Aristotelian seldom measured) the weight of the stone, the vertical height to which it had been raised, and the time required for it to achieve rest. Together with the resistance of the medium, these were the conceptual categories deployed by Aristotelian science when dealing with a falling body.[14] Normal research guided by them could not have produced the laws that Galileo discovered. It could only—and by another route it did—lead to the series of crises from which Galileo's view of the swinging stone emerged. As a result of those crises and of other intellectual changes besides, Galileo saw the swinging stone quite differently. Archimedes' work on floating bodies made the medium non-essential; the impetus theory rendered the motion symmetrical and enduring; and Neoplatonism directed Galileo's attention to the motion's circu-

[13] [Jacques] Hadamard, *Subconscient intuition, et logique dans la recherche scientifique* (*Conférence faite au Palais de la Découverte le 8 Décembre 1945* [Alençon, n.d.]), pp. 7–8. A much fuller account, though one exclusively restricted to mathematical innovations, is the same author's *The Psychology of Invention in the Mathematical Field* (Princeton, 1949).

[14] T. S. Kuhn, "A Function for Thought Experiments," in *Mélanges Alexandre Koyré*, ed. R. Taton and I. B. Cohen, to be published by Hermann (Paris) in 1963.

lar form.[15] He therefore measured only weight, radius, angular displacement, and time per swing, which were precisely the data that could be interpreted to yield Galileo's laws for the pendulum. In the event, interpretation proved almost unnecessary. Given Galileo's paradigms, pendulum-like regularities were very nearly accessible to inspection. How else are we to account for Galileo's discovery that the bob's period is entirely independent of amplitude, a discovery that the normal science stemming from Galileo had to eradicate and that we are quite unable to document today. Regularities that could not have existed for an Aristotelian (and that are, in fact, nowhere precisely exemplified by nature) were consequences of immediate experience for the man who saw the swinging stone as Galileo did.

Perhaps that example is too fanciful since the Aristotelians recorded no discussions of swinging stones. On their paradigm it was an extraordinarily complex phenomenon. But the Aristotelians did discuss the simpler case, stones falling without uncommon constraints, and the same differences of vision are apparent there. Contemplating a falling stone, Aristotle saw a change of state rather than a process. For him the relevant measures of a motion were therefore total distance covered and total time elapsed, parameters which yield what we should now call not speed but average speed.[16] Similarly, because the stone was impelled by its nature to reach its final resting point, Aristotle saw the relevant distance parameter at any instant during the motion as the distance *to* the final end point rather than as that *from* the origin of motion.[17] Those conceptual parameters underlie and give sense to most of his well-known "laws of motion." Partly through the impetus paradigm, however, and partly through a doctrine known as the latitude of forms, scholastic criticism changed this way of viewing motion. A stone moved by impetus gained more and more of it while receding from its

[15] A. Koyré, *Etudes Galiléennes* (Paris, 1939), I, 46–51; and "Galileo and Plato," *Journal of the History of Ideas*, IV (1943), 400–428.

[16] Kuhn, "A Function for Thought Experiments," in *Mélanges Alexandre Koyré* (see n. 14 for full citation).

[17] Koyré, *Etudes . . .*, II, 7–11.

starting point; distance from rather than distance to therefore became the revelant parameter. In addition, Aristotle's notion of speed was bifurcated by the scholastics into concepts that soon after Galileo became our average speed and instantaneous speed. But when seen through the paradigm of which these conceptions were a part, the falling stone, like the pendulum, exhibited its governing laws almost on inspection. Galileo was not one of the first men to suggest that stones fall with a uniformly accelerated motion.[18] Furthermore, he had developed his theorem on this subject together with many of its consequences before he experimented with an inclined plane. That theorem was another one of the network of new regularities accessible to genius in the world determined jointly by nature and by the paradigms upon which Galileo and his contemporaries had been raised. Living in that world, Galileo could still, when he chose, explain why Aristotle had seen what he did. Nevertheless, the immediate content of Galileo's experience with falling stones was not what Aristotle's had been.

It is, of course, by no means clear that we need be so concerned with "immediate experience"—that is, with the perceptual features that a paradigm so highlights that they surrender their regularities almost upon inspection. Those features must obviously change with the scientist's commitments to paradigms, but they are far from what we ordinarily have in mind when we speak of the raw data or the brute experience from which scientific research is reputed to proceed. Perhaps immediate experience should be set aside as fluid, and we should discuss instead the concrete operations and measurements that the scientist performs in his laboratory. Or perhaps the analysis should be carried further still from the immediately given. It might, for example, be conducted in terms of some neutral observation-language, perhaps one designed to conform to the retinal imprints that mediate what the scientist sees. Only in one of these ways can we hope to retrieve a realm in which experience is again stable once and for all—in which the pendulum and constrained fall are not different perceptions but rather

[18] Clagett, *op. cit.*, chaps. iv, vi, and ix.

different interpretations of the unequivocal data provided by observation of a swinging stone.

But is sensory experience fixed and neutral? Are theories simply man-made interpretations of given data? The epistemological viewpoint that has most often guided Western philosophy for three centuries dictates an immediate and unequivocal, Yes! In the absence of a developed alternative, I find it impossible to relinquish entirely that viewpoint. Yet it no longer functions effectively, and the attempts to make it do so through the introduction of a neutral language of observations now seem to me hopeless.

The operations and measurements that a scientist undertakes in the laboratory are not "the given" of experience but rather "the collected with difficulty." They are not what the scientist sees—at least not before his research is well advanced and his attention focused. Rather, they are concrete indices to the content of more elementary perceptions, and as such they are selected for the close scrutiny of normal research only because they promise opportunity for the fruitful elaboration of an accepted paradigm. Far more clearly than the immediate experience from which they in part derive, operations and measurements are paradigm-determined. Science does not deal in all possible laboratory manipulations. Instead, it selects those relevant to the juxtaposition of a paradigm with the immediate experience that that paradigm has partially determined. As a result, scientists with different paradigms engage in different concrete laboratory manipulations. The measurements to be performed on a pendulum are not the ones relevant to a case of constrained fall. Nor are the operations relevant for the elucidation of oxygen's properties uniformly the same as those required when investigating the characteristics of dephlogisticated air.

As for a pure observation-language, perhaps one will yet be devised. But three centuries after Descartes our hope for such an eventuality still depends exclusively upon a theory of perception and of the mind. And modern psychological experimentation is rapidly proliferating phenomena with which that theory can scarcely deal. The duck-rabbit shows that two men with the same retinal impressions can see different things; the inverting lenses show that two men with different retinal impressions can see the same thing. Psychology supplies a great deal of other evidence to the same effect, and the doubts that derive from it are readily reinforced by the history of attempts to exhibit an actual language of observation. No current attempt to achieve that end has yet come close to a generally applicable language of pure percepts. And those attempts that come closest share one characteristic that strongly reinforces several of this essay's main theses. From the start they presuppose a paradigm, taken either from a current scientific theory or from some fraction of everyday discourse, and they then try to eliminate from it all non-logical and non-perceptual terms. In a few realms of discourse this effort has been carried very far and with fascinating results. There can be no question that efforts of this sort are worth pursuing. But their result is a language that—like those employed in the sciences—embodies a host of expectations about nature and fails to function the moment these expectations are violated. Nelson Goodman makes exactly this point in describing the aims of his *Structure of Appearance:* "It is fortunate that nothing more [than phenomena known to exist] is in question; for the notion of 'possible' cases, of cases that do not exist but might have existed, is far from clear."[19] No language thus restricted to reporting a world fully known in advance can produce mere neutral and objective reports on "the given." Philosophical investigation has not yet provided even a hint of what a language able to do that would be like.

N.B.

Under these circumstances we may at least suspect that scientists are right in principle as well as in practice when they treat

[19] N. Goodman, *The Structure of Appearance* (Cambridge, Mass., 1951), pp. 4–5. The passage is worth quoting more extensively: "If all and only those residents of Wilmington in 1947 that weigh between 175 and 180 pounds have red hair, then 'red-haired 1947 resident of Wilmington' and '1947 resident of Wilmington weighing between 175 and 180 pounds' may be joined in a constructional definition. . . . The question whether there 'might have been' someone to whom one but not the other of these predicates would apply has no bearing . . . once we have determined that there is no such person. . . . It is fortunate that nothing more is in question; for the notion of 'possible' cases, of cases that do not exist but might have existed, is far from clear."

oxygen and pendulums (and perhaps also atoms and electrons) as the fundamental ingredients of their immediate experience. As a result of the paradigm-embodied experience of the race, the culture, and, finally, the profession, the world of the scientist has come to be populated with planets and pendulums, condensers and compound ores, and other such bodies besides. Compared with these objects of perception, both meter stick readings and retinal imprints are elaborate constructs to which experience has direct access only when the scientist, for the special purposes of his research, arranges that one or the other should do so. This is not to suggest that pendulums, for example, are the only things a scientist could possibly see when looking at a swinging stone. (We have already noted that members of another scientific community could see constrained fall.) But it is to suggest that the scientist who looks at a swinging stone can have no experience that is in principle more elementary than seeing a pendulum. The alternative is not some hypothetical "fixed" vision, but vision through another paradigm, one which makes the swinging stone something else.

All of this may seem more reasonable if we again remember that neither scientists nor laymen learn to see the world piecemeal or item by item. Except when all the conceptual and manipulative categories are prepared in advance—e.g., for the discovery of an additional transuranic element or for catching sight of a new house—both scientists and laymen sort out whole areas together from the flux of experience. The child who transfers the word 'mama' from all humans to all females and then to his mother is not just learning what 'mama' means or who his mother is. Simultaneously he is learning some of the differences between males and females as well as something about the ways in which all but one female will behave toward him. His reactions, expectations, and beliefs—indeed, much of his perceived world—change accordingly. By the same token, the Copernicans who denied its traditional title 'planet' to the sun were not only learning what 'planet' meant or what the sun was. Instead, they were changing the meaning of 'planet' so that it could continue to make useful distinctions in a world where all celestial bodies, not just the sun, were seen differently from the way they had been seen before. The same point could be made about any of our earlier examples. To see oxygen instead of dephlogisticated air, the condenser instead of the Leyden jar, or the pendulum instead of constrained fall, was only one part of an integrated shift in the scientist's vision of a great many related chemical, electrical, or dynamical phenomena. Paradigms determine large areas of experience at the same time.

It is, however, only after experience has been thus determined that the search for an operational definition or a pure observation-language can begin. The scientist or philosopher who asks what measurements or retinal imprints make the pendulum what it is must already be able to recognize a pendulum when he sees one. If he saw constrained fall instead, his question could not even be asked. And if he saw a pendulum, but saw it in the same way he saw a tuning fork or an oscillating balance, his question could not be answered. At least it could not be answered in the same way, because it would not be the same question. Therefore, though they are always legitimate and are occasionally extraordinarily fruitful, questions about retinal imprints or about the consequences of particular laboratory manipulations presuppose a world already perceptually and conceptually subdivided in a certain way. In a sense such questions are parts of normal science, for they depend upon the existence of a paradigm and they receive different answers as a result of paradigm change.

To conclude this section, let us henceforth neglect retinal impressions and again restrict attention to the laboratory operations that provide the scientist with concrete though fragmentary indices to what he has already seen. One way in which such laboratory operations change with paradigms has already been observed repeatedly. After a scientific revolution many old measurements and manipulations become irrelevant and are replaced by others instead. One does not apply all the same tests to oxygen as to dephlogisticated air. But changes of this sort are never total. Whatever he may then see, the scientist after a revolution is still looking at the same world. Furthermore, though he may previously have employed them differently, much of his language and most of his laboratory instruments are still the same as they were before. As a result, postrevolutionary science invariably includes many of the same manipulations, performed with the same instruments and described in the same terms, as its prerevolutionary predecessor. If these enduring manipulations have been changed at all, the change must lie either in their relation to the paradigm or in their concrete results. I now suggest, by the introduction of one last new example, that both these sorts of changes occur. Examining the work of Dalton and his contemporaries, we shall discover that one and the same operation, when it attaches to nature through a different paradigm, can become an index to a quite different aspect of nature's regularity. In addition, we shall see that occasionally the old manipulation in its new role will yield different concrete results.

Throughout much of the eighteenth century and into the nineteenth, European chemists almost universally believed that the elementary atoms of which all chemical species consisted were held together by forces of mutual affinity. Thus a lump of silver cohered because of the forces of affinity between silver corpuscles (until after Lavoisier these corpuscles were themselves thought of as compounded from still more elementary particles). On the same theory silver dissolved in acid (or salt in water) because the particles of acid attracted those of silver (or the particles of water attracted those of salt) more strongly than particles of these solutes attracted each other. Or again, copper would dissolve in the silver solution and precipitate silver, because the copper-acid affinity was greater than the affinity of acid for silver. A great many other phenomena were explained in the same way. In the eighteenth century the theory of elective affinity was an admirable chemical paradigm, widely and sometimes fruitfully deployed in the design and analysis of chemical experimentation.[20]

Affinity theory, however, drew the line separating physical

[20] H. Metzger, *Newton, Stahl, Boerhaave et la doctrine chimique* (Paris, 1930), pp. 34–68.

mixtures from chemical compounds in a way that has become unfamiliar since the assimilation of Dalton's work. Eighteenth-century chemists did recognize two sorts of processes. When mixing produced heat, light, effervescence or something else of the sort, chemical union was seen to have taken place. If, on the other hand, the particles in the mixture could be distinguished by eye or mechanically separated, there was only physical mixture. But in the very large number of intermediate cases—salt in water, alloys, glass, oxygen in the atmosphere, and so on—these crude criteria were of little use. Guided by their paradigm, most chemists viewed this entire intermediate range as chemical, because the processes of which it consisted were all governed by forces of the same sort. Salt in water or oxygen in nitrogen was just as much an example of chemical combination as was the combination produced by oxidizing copper. The arguments for viewing solutions as compounds were very strong. Affinity theory itself was well attested. Besides, the formation of a compound accounted for a solution's observed homogeneity. If, for example, oxygen and nitrogen were only mixed and not combined in the atmosphere, then the heavier gas, oxygen, should settle to the bottom. Dalton, who took the atmosphere to be a mixture, was never satisfactorily able to explain oxygen's failure to do so. The assimilation of his atomic theory ultimately created an anomaly where there had been none before.[21]

One is tempted to say that the chemists who viewed solutions as compounds differed from their successors only over a matter of definition. In one sense that may have been the case. But that sense is not the one that makes definitions mere conventional conveniences. In the eighteenth century mixtures were not fully distinguished from compounds by operational tests, and perhaps they could not have been. Even if chemists had looked for such tests, they would have sought criteria that made the solution a compound. The mixture-compound distinction was part of their paradigm—part of the way they viewed their whole

[21] *Ibid.*, pp. 124–29, 139–48. For Dalton, see Leonard K. Nash, *The Atomic-Molecular Theory* ("Harvard Case Histories in Experimental Science," Case 4; Cambridge, Mass., 1950), pp. 14–21.

field of research—and as such it was prior to any particular laboratory test, though not to the accumulated experience of chemistry as a whole.

But while chemistry was viewed in this way, chemical phenomena exemplified laws different from those that emerged with the assimilation of Dalton's new paradigm. In particular, while solutions remained compounds, no amount of chemical experimentation could by itself have produced the law of fixed proportions. At the end of the eighteenth century it was widely known that *some* compounds ordinarily contained fixed proportions by weight of their constituents. For some categories of reactions the German chemist Richter had even noted the further regularities now embraced by the law of chemical equivalents.[22] But no chemist made use of these regularities except in recipes, and no one until almost the end of the century thought of generalizing them. Given the obvious counterinstances, like glass or like salt in water, no generalization was possible without an abandonment of affinity theory and a reconceptualization of the boundaries of the chemist's domain. That consequence became explicit at the very end of the century in a famous debate between the French chemists Proust and Berthollet. The first claimed that all chemical reactions occurred in fixed proportion, the latter that they did not. Each collected impressive experimental evidence for his view. Nevertheless, the two men necessarily talked through each other, and their debate was entirely inconclusive. Where Berthollet saw a compound that could vary in proportion, Proust saw only a physical mixture.[23] To that issue neither experiment nor a change of definitional convention could be relevant. The two men were as fundamentally at cross-purposes as Galileo and Aristotle had been.

This was the situation during the years when John Dalton undertook the investigations that led finally to his famous chemical atomic theory. But until the very last stages of those investiga-

[22] J. R. Partington, *A Short History of Chemistry* (2d ed.; London, 1951), pp. 161–63.

[23] A. N. Meldrum, "The Development of the Atomic Theory: (1) Berthollet's Doctrine of Variable Proportions," *Manchester Memoirs*, LIV (1910), 1–16.

tions, Dalton was neither a chemist nor interested in chemistry. Instead, he was a meteorologist investigating the, for him, physical problems of the absorption of gases by water and of water by the atmosphere. Partly because his training was in a different specialty and partly because of his own work in that specialty, he approached these problems with a paradigm different from that of contemporary chemists. In particular, he viewed the mixture of gases or the absorption of a gas in water as a physical process, one in which forces of affinity played no part. To him, therefore, the observed homogeneity of solutions was a problem, but one which he thought he could solve if he could determine the relative sizes and weights of the various atomic particles in his experimental mixtures. It was to determine these sizes and weights that Dalton finally turned to chemistry, supposing from the start that, in the restricted range of reactions that he took to be chemical, atoms could only combine one-to-one or in some other simple whole-number ratio.[24] That natural assumption did enable him to determine the sizes and weights of elementary particles, but it also made the law of constant proportion a tautology. For Dalton, any reaction in which the ingredients did not enter in fixed proportion was *ipso facto* not a purely chemical process. A law that experiment could not have established before Dalton's work, became, once that work was accepted, a constitutive principle that no single set of chemical measurements could have upset. As a result of what is perhaps our fullest example of a scientific revolution, the same chemical manipulations assumed a relationship to chemical generalization very different from the one they had had before.

Needless to say, Dalton's conclusions were widely attacked when first announced. Berthollet, in particular, was never convinced. Considering the nature of the issue, he need not have been. But to most chemists Dalton's new paradigm proved convincing where Proust's had not been, for it had implications far wider and more important than a new criterion for distinguish-

[24] L. K. Nash, "The Origin of Dalton's Chemical Atomic Theory," *Isis,* XLVII (1956), 101–16.

ing a mixture from a compound. If, for example, atoms could combine chemically only in simple whole-number ratios, then a re-examination of existing chemical data should disclose examples of multiple as well as of fixed proportions. Chemists stopped writing that the two oxides of, say, carbon contained 56 per cent and 72 per cent of oxygen by weight; instead they wrote that one weight of carbon would combine either with 1.3 or with 2.6 weights of oxygen. When the results of old manipulations were recorded in this way, a 2:1 ratio leaped to the eye; and this occurred in the analysis of many well-known reactions and of new ones besides. In addition, Dalton's paradigm made it possible to assimilate Richter's work and to see its full generality. Also, it suggested new experiments, particularly those of Gay-Lussac on combining volumes, and these yielded still other regularities, ones that chemists had not previously dreamed of. What chemists took from Dalton was not new experimental laws but a new way of practicing chemistry (he himself called it the "new system of chemical philosophy"), and this proved so rapidly fruitful that only a few of the older chemists in France and Britain were able to resist it.[25] As a result, chemists came to live in a world where reactions behaved quite differently from the way they had before.

As all this went on, one other typical and very important change occurred. Here and there the very numerical data of chemistry began to shift. When Dalton first searched the chemical literature for data to support his physical theory, he found some records of reactions that fitted, but he can scarcely have avoided finding others that did not. Proust's own measurements on the two oxides of copper yielded, for example, an oxygen weight-ratio of 1.47:1 rather than the 2:1 demanded by the atomic theory; and Proust is just the man who might have been expected to achieve the Daltonian ratio.[26] He was, that is, a fine

[25] A. N. Meldrum, "The Development of the Atomic Theory: (6) The Reception Accorded to the Theory Advocated by Dalton," *Manchester Memoirs*, LV (1911), 1–10.

[26] For Proust, see Meldrum, "Berthollet's Doctrine of Variable Proportions," *Manchester Memoirs*, LIV (1910), 8. The detailed history of the gradual changes in measurements of chemical composition and of atomic weights has yet to be written, but Partington, *op. cit.*, provides many useful leads to it.

experimentalist, and his view of the relation between mixtures and compounds was very close to Dalton's. But it is hard to make nature fit a paradigm. That is why the puzzles of normal science are so challenging and also why measurements undertaken without a paradigm so seldom lead to any conclusions at all. Chemists could not, therefore, simply accept Dalton's theory on the evidence, for much of that was still negative. Instead, even after accepting the theory, they had still to beat nature into line, a process which, in the event, took almost another generation. When it was done, even the percentage composition of well-known compounds was different. The data themselves had changed. That is the last of the senses in which we may want to say that after a revolution scientists work in a different world.

XI. The Invisibility of Revolutions

We must still ask how scientific revolutions close. Before doing so, however, a last attempt to reinforce conviction about their existence and nature seems called for. I have so far tried to display revolutions by illustration, and the examples could be multiplied *ad nauseam*. But clearly, most of them, which were deliberately selected for their familiarity, have customarily been viewed not as revolutions but as additions to scientific knowledge. That same view could equally well be taken of any additional illustrations, and these would probably be ineffective. I suggest that there are excellent reasons why revolutions have proved to be so nearly invisible. Both scientists and laymen take much of their image of creative scientific activity from an authoritative source that systematically disguises—partly for important functional reasons—the existence and significance of scientific revolutions. Only when the nature of that authority is recognized and analyzed can one hope to make historical example fully effective. Furthermore, though the point can be fully developed only in my concluding section, the analysis now required will begin to indicate one of the aspects of scientific work that most clearly distinguishes it from every other creative pursuit except perhaps theology.

As the source of authority, I have in mind principally textbooks of science together with both the popularizations and the philosophical works modeled on them. All three of these categories—until recently no other significant sources of information about science have been available except through the practice of research—have one thing in common. They address themselves to an already articulated body of problems, data, and theory, most often to the particular set of paradigms to which the scientific community is committed at the time they are written. Textbooks themselves aim to communicate the vocabulary and syntax of a contemporary scientific language. Popularizations attempt to describe these same applications in a language closer to that of everyday life. And philosophy of science, particularly that of the English-speaking world, analyzes the logical structure of the same completed body of scientific knowledge. Though a fuller treatment would necessarily deal with the very real distinctions between these three genres, it is their similarities that most concern us here. All three record the stable *outcome* of past revolutions and thus display the bases of the current normal-scientific tradition. To fulfill their function they need not provide authentic information about the way in which those bases were first recognized and then embraced by the profession. In the case of textbooks, at least, there are even good reasons why, in these matters, they should be systematically misleading.

We noted in Section II that an increasing reliance on textbooks or their equivalent was an invariable concomitant of the emergence of a first paradigm in any field of science. The concluding section of this essay will argue that the domination of a mature science by such texts significantly differentiates its developmental pattern from that of other fields. For the moment let us simply take it for granted that, to an extent unprecedented in other fields, both the layman's and the practitioner's knowledge of science is based on textbooks and a few other types of literature derived from them. Textbooks, however, being pedagogic vehicles for the perpetuation of normal science, have to be rewritten in whole or in part whenever the language, problem-structure, or standards of normal science change. In short, they have to be rewritten in the aftermath of each scientific revolution, and, once rewritten, they inevitably disguise not only the role but the very existence of the revolutions that produced them. Unless he has personally experienced a revolution in his own lifetime, the historical sense either of the working scientist or of the lay reader of textbook literature extends only to the outcome of the most recent revolutions in the field.

Textbooks thus begin by truncating the scientist's sense of his discipline's history and then proceed to supply a substitute for what they have eliminated. Characteristically, textbooks of science contain just a bit of history, either in an introductory chapter or, more often, in scattered references to the great heroes of an earlier age. From such references both students and professionals come to feel like participants in a long-standing historical tradition. Yet the textbook-derived tradition in which scientists come to sense their participation is one that, in fact, never existed. For reasons that are both obvious and highly functional, science textbooks (and too many of the older histories of science) refer only to that part of the work of past scientists that can easily be viewed as contributions to the statement and solution of the texts' paradigm problems. Partly by selection and partly by distortion, the scientists of earlier ages are implicitly represented as having worked upon the same set of fixed problems and in accordance with the same set of fixed canons that the most recent revolution in scientific theory and method has made seem scientific. No wonder that textbooks and the historical tradition they imply have to be rewritten after each scientific revolution. And no wonder that, as they are rewritten, science once again comes to seem largely cumulative.

Scientists are not, of course, the only group that tends to see its discipline's past developing linearly toward its present vantage. The temptation to write history backward is both omnipresent and perennial. But scientists are more affected by the temptation to rewrite history, partly because the results of scientific research show no obvious dependence upon the historical context of the inquiry, and partly because, except during crisis and revolution, the scientist's contemporary position seems so secure. More historical detail, whether of science's present or of its past, or more responsibility to the historical details that are presented, could only give artificial status to human idiosyncrasy, error, and confusion. Why dignify what science's best and most persistent efforts have made it possible to discard? The depreciation of historical fact is deeply, and probably functionally, ingrained in the ideology of the scientific profession, the same profession that places the highest of all values upon factual details of other sorts. Whitehead caught the unhistorical spirit of the scientific community when he wrote, "A science that hesitates to forget its founders is lost." Yet he was not quite right, for the sciences, like other professional enterprises, do need their heroes and do preserve their names. Fortunately, instead of forgetting these heroes, scientists have been able to forget or revise their works.

The result is a persistent tendency to make the history of science look linear or cumulative, a tendency that even affects scientists looking back at their own research. For example, all three of Dalton's incompatible accounts of the development of his chemical atomism make it appear that he was interested from an early date in just those chemical problems of combining proportions that he was later famous for having solved. Actually those problems seem only to have occurred to him with their solutions, and then not until his own creative work was very nearly complete.[1] What all of Dalton's accounts omit are the revolutionary effects of applying to chemistry a set of questions and concepts previously restricted to physics and meteorology. That is what Dalton did, and the result was a reorientation toward the field, a reorientation that taught chemists to ask new questions about and to draw new conclusions from old data.

Or again, Newton wrote that Galileo had discovered that the constant force of gravity produces a motion proportional to the square of the time. In fact, Galileo's kinematic theorem does take that form when embedded in the matrix of Newton's own dynamical concepts. But Galileo said nothing of the sort. His discussion of falling bodies rarely alludes to forces, much less to a uniform gravitational force that causes bodies to fall.[2] By crediting to Galileo the answer to a question that Galileo's paradigms did not permit to be asked, Newton's account hides the effect of a small but revolutionary reformulation in the questions that scientists asked about motion as well as in the

[1] L. K. Nash, "The Origins of Dalton's Chemical Atomic Theory," *Isis*, XLVII (1956), 101–16.

[2] For Newton's remark, see Florian Cajori (ed.), *Sir Isaac Newton's Mathematical Principles of Natural Philosophy and His System of the World* (Berkeley, Calif., 1946), p. 21. The passage should be compared with Galileo's own discussion in his *Dialogues concerning Two New Sciences*, trans. H. Crew and A. de Salvio (Evanston, Ill., 1946), pp. 154–76.

answers they felt able to accept. But it is just this sort of change in the formulation of questions and answers that accounts, far more than novel empirical discoveries, for the transition from Aristotelian to Galilean and from Galilean to Newtonian dynamics. By disguising such changes, the textbook tendency to make the development of science linear hides a process that lies at the heart of the most significant episodes of scientific development.

The preceding examples display, each within the context of a single revolution, the beginnings of a reconstruction of history that is regularly completed by postrevolutionary science texts. But in that completion more is involved than a multiplication of the historical misconstructions illustrated above. Those misconstructions render revolutions invisible; the arrangement of the still visible material in science texts implies a process that, if it existed, would deny revolutions a function. Because they aim quickly to acquaint the student with what the contemporary scientific community thinks it knows, textbooks treat the various experiments, concepts, laws, and theories of the current normal science as separately and as nearly seriatim as possible. As pedagogy this technique of presentation is unexceptionable. But when combined with the generally unhistorical air of science writing and with the occasional systematic misconstructions discussed above, one strong impression is overwhelmingly likely to follow: science has reached its present state by a series of individual discoveries and inventions that, when gathered together, constitute the modern body of technical knowledge. From the beginning of the scientific enterprise, a textbook presentation implies, scientists have striven for the particular objectives that are embodied in today's paradigms. One by one, in a process often compared to the addition of bricks to a building, scientists have added another fact, concept, law, or theory to the body of information supplied in the contemporary science text.

But that is not the way a science develops. Many of the puzzles of contemporary normal science did not exist until after the most recent scientific revolution. Very few of them can be traced back to the historic beginning of the science within which they now occur. Earlier generations pursued their own problems with their own instruments and their own canons of solution. Nor is it just the problems that have changed. Rather the whole network of fact and theory that the textbook paradigm fits to nature has shifted. Is the constancy of chemical composition, for example, a mere fact of experience that chemists could have discovered by experiment within any one of the worlds within which chemists have practiced? Or is it rather one element—and an indubitable one, at that—in a new fabric of associated fact and theory that Dalton fitted to the earlier chemical experience as a whole, changing that experience in the process? Or by the same token, is the constant acceleration produced by a constant force a mere fact that students of dynamics have always sought, or is it rather the answer to a question that first arose only within Newtonian theory and that that theory could answer from the body of information available before the question was asked?

These questions are here asked about what appear as the piecemeal-discovered facts of a textbook presentation. But obviously, they have implications as well for what the text presents as theories. Those theories, of course, do "fit the facts," but only by transforming previously accessible information into facts that, for the preceding paradigm, had not existed at all. And that means that theories too do not evolve piecemeal to fit facts that were there all the time. Rather, they emerge together with the facts they fit from a revolutionary reformulation of the preceding scientific tradition, a tradition within which the knowledge-mediated relationship between the scientist and nature was not quite the same.

One last example may clarify this account of the impact of textbook presentation upon our image of scientific development. Every elementary chemistry text must discuss the concept of a chemical element. Almost always, when that notion is introduced, its origin is attributed to the seventeenth-century chemist, Robert Boyle, in whose *Sceptical Chymist* the attentive reader will find a definition of 'element' quite close to that in use today. Reference to Boyle's contribution helps to make the neophyte aware that chemistry did not begin with the sulfa drugs; in addition, it tells him that one of the scientist's traditional tasks is to invent concepts of this sort. As a part of the pedagogic arsenal that makes a man a scientist, the attribution is immensely successful. Nevertheless, it illustrates once more the pattern of historical mistakes that misleads both students and laymen about the nature of the scientific enterprise.

According to Boyle, who was quite right, his "definition" of an element was no more than a paraphrase of a traditional chemical concept; Boyle offered it only in order to argue that no such thing as a chemical element exists; as history, the textbook version of Boyle's contribution is quite mistaken.[3] That mistake, of course, is trivial, though no more so than any other misrepresentation of data. What is not trivial, however, is the impression of science fostered when this sort of mistake is first compounded and then built into the technical structure of the text. Like 'time,' 'energy,' 'force,' or 'particle,' the concept of an element is the sort of textbook ingredient that is often not invented or discovered at all. Boyle's definition, in particular, can be traced back at least to Aristotle and forward through Lavoisier into modern texts. Yet that is not to say that science has possessed the modern concept of an element since antiquity. Verbal definitions like Boyle's have little scientific content when considered by themselves. They are not full logical specifications of meaning (if there are such), but more nearly pedagogic aids. The scientific concepts to which they point gain full significance only when related, within a text or other systematic presentation, to other scientific concepts, to manipulative procedures, and to paradigm applications. It follows that concepts like that of an element can scarcely be invented independent of context. Furthermore, given the context, they rarely require invention because they are already at hand. Both Boyle and Lavoisier changed the chemical significance of 'element' in important ways. But they did not invent the notion

[3] T. S. Kuhn, "Robert Boyle and Structural Chemistry in the Seventeenth Century," *Isis*, XLIII (1952), 26–29.

or even change the verbal formula that serves as its definition. Nor, as we have seen, did Einstein have to invent or even explicitly redefine 'space' and 'time' in order to give them new meaning within the context of his work.

What then was Boyle's historical function in that part of his work that includes the famous "definition"? He was a leader of a scientific revolution that, by changing the relation of 'element' to chemical manipulation and chemical theory, transformed the notion into a tool quite different from what it had been before and transformed both chemistry and the chemist's world in the process.[4] Other revolutions, including the one that centers around Lavoisier, were required to give the concept its modern form and function. But Boyle provides a typical example both of the process involved at each of these stages and of what happens to that process when existing knowledge is embodied in a textbook. More than any other single aspect of science, that pedagogic form has determined our image of the nature of science and of the role of discovery and invention in its advance.

[4] Marie Boas, in her *Robert Boyle and Seventeenth-Century Chemistry* (Cambridge, 1958), deals in many places with Boyle's positive contributions to the evolution of the concept of a chemical element.

XII. The Resolution of Revolutions

The textbooks we have just been discussing are produced only in the aftermath of a scientific revolution. They are the bases for a new tradition of normal science. In taking up the question of their structure we have clearly missed a step. What is the process by which a new candidate for paradigm replaces its predecessor? Any new interpretation of nature, whether a discovery or a theory, emerges first in the mind of one or a few individuals. It is they who first learn to see science and the world differently, and their ability to make the transition is facilitated by two circumstances that are not common to most other members of their profession. Invariably their attention has been intensely concentrated upon the crisis-provoking problems; usually, in addition, they are men so young or so new to the crisis-ridden field that practice has committed them less deeply than most of their contemporaries to the world view and rules determined by the old paradigm. How are they able, what must they do, to convert the entire profession or the relevant professional subgroup to their way of seeing science and the world? What causes the group to abandon one tradition of normal research in favor of another?

To see the urgency of those questions, remember that they are the only reconstructions the historian can supply for the philosopher's inquiry about the testing, verification, or falsification of established scientific theories. In so far as he is engaged in normal science, the research worker is a solver of puzzles, not a tester of paradigms. Though he may, during the search for a particular puzzle's solution, try out a number of alternative approaches, rejecting those that fail to yield the desired result, he is not testing the *paradigm* when he does so. Instead he is like the chess player who, with a problem stated and the board physically or mentally before him, tries out various alternative moves in the search for a solution. These trial attempts, whether by the chess player or by the scientist, are trials only of themselves, not of the rules of the game. They are possible only so long as the paradigm itself is taken for granted. Therefore, paradigm-testing occurs only after persistent failure to solve a noteworthy puzzle has given rise to crisis. And even then it occurs only after the sense of crisis has evoked an alternate candidate for paradigm. In the sciences the testing situation never consists, as puzzle-solving does, simply in the comparison of a single paradigm with nature. Instead, testing occurs as part of the competition between two rival paradigms for the allegiance of the scientific community.

Closely examined, this formulation displays unexpected and probably significant parallels to two of the most popular contemporary philosophical theories about verification. Few philosophers of science still seek absolute criteria for the verification of scientific theories. Noting that no theory can ever be exposed to all possible relevant tests, they ask not whether a theory has been verified but rather about its probability in the light of the evidence that actually exists. And to answer that question one important school is driven to compare the ability of different theories to explain the evidence at hand. That insistence on comparing theories also characterizes the historical situation in which a new theory is accepted. Very probably it points one of the directions in which future discussions of verification should go.

In their most usual forms, however, probabilistic verification theories all have recourse to one or another of the pure or neutral observation-languages discussed in Section X. One probabilistic theory asks that we compare the given scientific theory with all others that might be imagined to fit the same collection of observed data. Another demands the construction in imagination of all the tests that the given scientific theory might conceivably be asked to pass.[1] Apparently some such construction is necessary for the computation of specific probabilities, absolute or relative, and it is hard to see how such a construction can

[1] For a brief sketch of the main routes to probabilistic verification theories, see Ernest Nagel, *Principles of the Theory of Probability*, Vol. I, No. 6, of *International Encyclopedia of Unified Science*, pp. 60–75.

possibly be achieved. If, as I have already urged, there can be no scientifically or empirically neutral system of language or concepts, then the proposed construction of alternate tests and theories must proceed from within one or another paradigm-based tradition. Thus restricted it would have no access to all possible experiences or to all possible theories. As a result, probabilistic theories disguise the verification situation as much as they illuminate it. Though that situation does, as they insist, depend upon the comparison of theories and of much widespread evidence, the theories and observations at issue are always closely related to ones already in existence. Verification is like natural selection: it picks out the most viable among the actual alternatives in a particular historical situation. Whether that choice is the best that could have been made if still other alternatives had been available or if the data had been of another sort is not a question that can usefully be asked. There are no tools to employ in seeking answers to it.

A very different approach to this whole network of problems has been developed by Karl R. Popper who denies the existence of any verification procedures at all.[2] Instead, he emphasizes the importance of falsification, i.e., of the test that, because its outcome is negative, necessitates the rejection of an established theory. Clearly, the role thus attributed to falsification is much like the one this essay assigns to anomalous experiences, i.e., to experiences that, by evoking crisis, prepare the way for a new theory. Nevertheless, anomalous experiences may not be identified with falsifying ones. Indeed, I doubt that the latter exist. As has repeatedly been emphasized before, no theory ever solves all the puzzles with which it is confronted at a given time; nor are the solutions already achieved often perfect. On the contrary, it is just the incompleteness and imperfection of the existing data-theory fit that, at any time, define many of the puzzles that characterize normal science. If any and every failure to fit were ground for theory rejection, all theories ought to be rejected at all times. On the other hand, if only severe failure

[2] K. R. Popper, *The Logic of Scientific Discovery* (New York, 1959), esp. chaps. i–iv.

to fit justifies theory rejection, then the Popperians will require some criterion of "improbability" or of "degree of falsification." In developing one they will almost certainly encounter the same network of difficulties that has haunted the advocates of the various probabilistic verification theories.

Many of the preceding difficulties can be avoided by recognizing that both of these prevalent and opposed views about the underlying logic of scientific inquiry have tried to compress two largely separate processes into one. Popper's anomalous experience is important to science because it evokes competitors for an existing paradigm. But falsification, though it surely occurs, does not happen with, or simply because of, the emergence of an anomaly or falsifying instance. Instead, it is a subsequent and separate process that might equally well be called verification since it consists in the triumph of a new paradigm over the old one. Furthermore, it is in that joint verification-falsification process that the probabilist's comparison of theories plays a central role. Such a two-stage formulation has, I think, the virtue of great verisimilitude, and it may also enable us to begin explicating the rôle of agreement (or disagreement) between fact and theory in the verification process. To the historian, at least, it makes little sense to suggest that verification is establishing the agreement of fact with theory. All historically significant theories have agreed with the facts, but only more or less. There is no more precise answer to the question whether or how well an individual theory fits the facts. But questions much like that can be asked when theories are taken collectively or even in pairs. It makes a great deal of sense to ask which of two actual and competing theories fits the facts *better*. Though neither Priestley's nor Lavoisier's theory, for example, agreed precisely with existing observations, few contemporaries hesitated more than a decade in concluding that Lavoisier's theory provided the better fit of the two.

This formulation, however, makes the task of choosing between paradigms look both easier and more familiar than it is. If there were but one set of scientific problems, one world within which to work on them, and one set of standards for their solution, paradigm competition might be settled more or less routinely by some process like counting the number of problems solved by each. But, in fact, these conditions are never met completely. The proponents of competing paradigms are always at least slightly at cross-purposes. Neither side will grant all the non-empirical assumptions that the other needs in order to make its case. Like Proust and Berthollet arguing about the composition of chemical compounds, they are bound partly to talk through each other. Though each may hope to convert the other to his way of seeing his science and its problems, neither may hope to prove his case. The competition between paradigms is not the sort of battle that can be resolved by proofs.

We have already seen several reasons why the proponents of competing paradigms must fail to make complete contact with each other's viewpoints. Collectively these reasons have been described as the incommensurability of the pre- and postrevolutionary normal-scientific traditions, and we need only recapitulate them briefly here. In the first place, the proponents of competing paradigms will often disagree about the list of problems that any candidate for paradigm must resolve. Their standards or their definitions of science are not the same. Must a theory of motion explain the cause of the attractive forces between particles of matter or may it simply note the existence of such forces? Newton's dynamics was widely rejected because, unlike both Aristotle's and Descartes's theories, it implied the latter answer to the question. When Newton's theory had been accepted, a question was therefore banished from science. That question, however, was one that general relativity may proudly claim to have solved. Or again, as disseminated in the nineteenth century, Lavoisier's chemical theory inhibited chemists from asking why the metals were so much alike, a question that phlogistic chemistry had both asked and answered. The transition to Lavoisier's paradigm had, like the transition to Newton's, meant a loss not only of a permissible question but of an achieved solution. That loss was not, however, permanent either. In the twentieth century questions about the qualities of chemical substances have entered science again, together with some answers to them.

More is involved, however, than the incommensurability of standards. Since new paradigms are born from old ones, they ordinarily incorporate much of the vocabulary and apparatus, both conceptual and manipulative, that the traditional paradigm had previously employed. But they seldom employ these borrowed elements in quite the traditional way. Within the new paradigm, old terms, concepts, and experiments fall into new relationships one with the other. The inevitable result is what we must call, though the term is not quite right, a misunderstanding between the two competing schools. The laymen who scoffed at Einstein's general theory of relativity because space could not be "curved"—it was not that sort of thing—were not simply wrong or mistaken. Nor were the mathematicians, physicists, and philosophers who tried to develop a Euclidean version of Einstein's theory.[3] What had previously been meant by space was necessarily flat, homogeneous, isotropic, and unaffected by the presence of matter. If it had not been, Newtonian physics would not have worked. To make the transition to Einstein's universe, the whole conceptual web whose strands are space, time, matter, force, and so on, had to be shifted and laid down again on nature whole. Only men who had together undergone or failed to undergo that transformation would be able to discover precisely what they agreed or disagreed about. Communication across the revolutionary divide is inevitably partial. Consider, for another example, the men who called Copernicus mad because he proclaimed that the earth moved. They were not either just wrong or quite wrong. Part of what they meant by 'earth' was fixed position. Their earth, at least, could not be moved. Correspondingly, Copernicus' innovation was not simply to move the earth. Rather, it was a whole new way of regarding the problems of physics and astronomy,

[3] For lay reactions to the concept of curved space, see Philipp Frank, *Einstein, His Life and Times,* trans. and ed. G. Rosen and S. Kusaka (New York, 1947), pp. 142–46. For a few of the attempts to preserve the gains of general relativity within a Euclidean space, see C. Nordmann, *Einstein and the Universe,* trans. J. McCabe (New York, 1922), chap. ix.

one that necessarily changed the meaning of both 'earth' and 'motion.'[4] Without those changes the concept of a moving earth was mad. On the other hand, once they had been made and understood, both Descartes and Huyghens could realize that the earth's motion was a question with no content for science.[5]

These examples point to the third and most fundamental aspect of the incommensurability of competing paradigms. In a sense that I am unable to explicate further, the proponents of competing paradigms practice their trades in different worlds. One contains constrained bodies that fall slowly, the other pendulums that repeat their motions again and again. In one, solutions are compounds, in the other mixtures. One is embedded in a flat, the other in a curved, matrix of space. Practicing in different worlds, the two groups of scientists see different things when they look from the same point in the same direction. Again, that is not to say that they can see anything they please. Both are looking at the world, and what they look at has not changed. But in some areas they see different things, and they see them in different relations one to the other. That is why a law that cannot even be demonstrated to one group of scientists may occasionally seem intuitively obvious to another. Equally, it is why, before they can hope to communicate fully, one group or the other must experience the conversion that we have been calling a paradigm shift. Just because it is a transition between incommensurables, the transition between competing paradigms cannot be made a step at a time, forced by logic and neutral experience. Like the gestalt switch, it must occur all at once (though not necessarily in an instant) or not at all.

How, then, are scientists brought to make this transposition? Part of the answer is that they are very often not. Copernicanism made few converts for almost a century after Copernicus' death. Newton's work was not generally accepted, particularly on the Continent, for more than half a century after the *Prin-*

[4] T. S. Kuhn, *The Copernican Revolution* (Cambridge, Mass., 1957), chaps. iii, iv, and vii. The extent to which heliocentrism was more than a strictly astronomical issue is a major theme of the entire book.

[5] Max Jammer, *Concepts of Space* (Cambridge, Mass., 1954), pp. 118–24.

cipia appeared.[6] Priestley never accepted the oxygen theory, nor Lord Kelvin the electromagnetic theory, and so on. The difficulties of conversion have often been noted by scientists themselves. Darwin, in a particularly perceptive passage at the end of his *Origin of Species,* wrote: "Although I am fully convinced of the truth of the views given in this volume . . . , I by no means expect to convince experienced naturalists whose minds are stocked with a multitude of facts all viewed, during a long course of years, from a point of view directly opposite to mine. . . . [B]ut I look with confidence to the future,—to young and rising naturalists, who will be able to view both sides of the question with impartiality."[7] And Max Planck, surveying his own career in his *Scientific Autobiography,* sadly remarked that "a new scientific truth does not triumph by convincing its opponents and making them see the light, but rather because its opponents eventually die, and a new generation grows up that is familiar with it."[8]

These facts and others like them are too commonly known to need further emphasis. But they do need re-evaluation. In the past they have most often been taken to indicate that scientists, being only human, cannot always admit their errors, even when confronted with strict proof. I would argue, rather, that in these matters neither proof nor error is at issue. The transfer of allegiance fom paradigm to paradigm is a conversion experience that cannot be forced. Lifelong resistance, particularly from those whose productive careers have committed them to an older tradition of normal science, is not a violation of scientific standards but an index to the nature of scientific research itself. The source of resistance is the assurance that the older paradigm will ultimately solve all its problems, that nature can be shoved

[6] I. B. Cohen, *Franklin and Newton: An Inquiry into Speculative Newtonian Experimental Science and Franklin's Work in Electricity as an Example Thereof* (Philadelphia, 1956), pp. 93–94.

[7] Charles Darwin, *On the Origin of Species . . .* (authorized edition from 6th English ed.; New York, 1889), II, 295–96.

[8] Max Planck, *Scientific Autobiography and Other Papers,* trans. F. Gaynor (New York, 1949), pp. 33–34.

into the box the paradigm provides. Inevitably, at times of revolution, that assurance seems stubborn and pigheaded as indeed it sometimes becomes. But it is also something more. That same assurance is what makes normal or puzzle-solving science possible. And it is only through normal science that the professional community of scientists succeeds, first, in exploiting the potential scope and precision of the older paradigm and, then, in isolating the difficulty through the study of which a new paradigm may emerge.

Still, to say that resistance is inevitable and legitimate, that paradigm change cannot be justified by proof, is not to say that no arguments are relevant or that scientists cannot be persuaded to change their minds. Though a generation is sometimes required to effect the change, scientific communities have again and again been converted to new paradigms. Furthermore, these conversions occur not despite the fact that scientists are human but because they are. Though some scientists, particularly the older and more experienced ones, may resist indefinitely, most of them can be reached in one way or another. Conversions will occur a few at a time until, after the last holdouts have died, the whole profession will again be practicing under a single, but now a different, paradigm. We must therefore ask how conversion is induced and how resisted.

What sort of answer to that question may we expect? Just because it is asked about techniques of persuasion, or about argument and counterargument in a situation in which there can be no proof, our question is a new one, demanding a sort of study that has not previously been undertaken. We shall have to settle for a very partial and impressionistic survey. In addition, what has already been said combines with the result of that survey to suggest that, when asked about persuasion rather than proof, the question of the nature of scientific argument has no single or uniform answer. Individual scientists embrace a new paradigm for all sorts of reasons and usually for several at once. Some of these reasons—for example, the sun worship that helped make Kepler a Copernican—lie outside the apparent sphere of science entirely.[9] Others must depend upon idiosyncrasies of autobiography and personality. Even the nationality or the prior reputation of the innovator and his teachers can sometimes play a significant role.[10] Ultimately, therefore, we must learn to ask this question differently. Our concern will not then be with the arguments that in fact convert one or another individual, but rather with the sort of community that always sooner or later re-forms as a single group. That problem, however, I postpone to the final section, examining meanwhile some of the sorts of argument that prove particularly effective in the battles over paradigm change.

Probably the single most prevalent claim advanced by the proponents of a new paradigm is that they can solve the problems that have led the old one to a crisis. When it can legitimately be made, this claim is often the most effective one possible. In the area for which it is advanced the paradigm is known to be in trouble. That trouble has repeatedly been explored, and attempts to remove it have again and again proved vain. "Crucial experiments"—those able to discriminate particularly sharply between the two paradigms—have been recognized and attested before the new paradigm was even invented. Copernicus thus claimed that he had solved the long-vexing problem of the length of the calendar year, Newton that he had reconciled terrestrial and celestial mechanics, Lavoisier that he had solved the problems of gas-identity and of weight relations, and Einstein that he had made electrodynamics compatible with a revised science of motion.

Claims of this sort are particularly likely to succeed if the new paradigm displays a quantitative precision strikingly better than

[9] For the role of sun worship in Kepler's thought, see E. A. Burtt, *The Metaphysical Foundations of Modern Physical Science* (rev. ed.; New York, 1932), pp. 44–49.

[10] For the role of reputation, consider the following: Lord Rayleigh, at a time when his reputation was established, submitted to the British Association a paper on some paradoxes of electrodynamics. His name was inadvertently omitted when the paper was first sent, and the paper itself was at first rejected as the work of some "paradoxer." Shortly afterwards, with the author's name in place, the paper was accepted with profuse apologies (R. J. Strutt, 4th Baron Rayleigh, *John William Strutt, Third Baron Rayleigh* [New York, 1924], p. 228).

its older competitor. The quantitative superiority of Kepler's Rudolphine tables to all those computed from the Ptolemaic theory was a major factor in the conversion of astronomers to Copernicanism. Newton's success in predicting quantitative astronomical observations was probably the single most important reason for his theory's triumph over its more reasonable but uniformly qualitative competitors. And in this century the striking quantitative success of both Planck's radiation law and the Bohr atom quickly persuaded many physicists to adopt them even though, viewing physical science as a whole, both these contributions created many more problems than they solved.[11]

The claim to have solved the crisis-provoking problems is, however, rarely sufficient by itself. Nor can it always legitimately be made. In fact, Copernicus' theory was not more accurate than Ptolemy's and did not lead directly to any improvement in the calendar. Or again, the wave theory of light was not, for some years after it was first announced, even as successful as its corpuscular rival in resolving the polarization effects that were a principal cause of the optical crisis. Sometimes the looser practice that characterizes extraordinary research will produce a candidate for paradigm that initially helps not at all with the problems that have evoked crisis. When that occurs, evidence must be drawn from other parts of the field as it often is anyway. In those other areas particularly persuasive arguments can be developed if the new paradigm permits the prediction of phenomena that had been entirely unsuspected while the old one prevailed.

Copernicus' theory, for example, suggested that planets should be like the earth, that Venus should show phases, and that the universe must be vastly larger than had previously been supposed. As a result, when sixty years after his death the telescope suddenly displayed mountains on the moon, the phases of Venus, and an immense number of previously unsuspected stars,

[11] For the problems created by the quantum theory, see F. Reiche, *The Quantum Theory* (London, 1922), chaps. ii, vi–ix. For the other examples in this paragraph, see the earlier references in this section.

those observations brought the new theory a great many converts, particularly among non-astronomers.[12] In the case of the wave theory, one main source of professional conversions was even more dramatic. French resistance collapsed suddenly and relatively completely when Fresnel was able to demonstrate the existence of a white spot at the center of the shadow of a circular disk. That was an effect that not even he had anticipated but that Poisson, initially one of his opponents, had shown to be a necessary if absurd consequence of Fresnel's theory.[13] Because of their shock value and because they have so obviously not been "built into" the new theory from the start, arguments like these prove especially persuasive. And sometimes that extra strength can be exploited even though the phenomenon in question had been observed long before the theory that accounts for it was first introduced. Einstein, for example, seems not to have anticipated that general relativity would account with precision for the well-known anomaly in the motion of Mercury's perihelion, and he experienced a corresponding triumph when it did so.[14]

All the arguments for a new paradigm discussed so far have been based upon the competitors' comparative ability to solve problems. To scientists those arguments are ordinarily the most significant and persuasive. The preceding examples should leave no doubt about the source of their immense appeal. But, for reasons to which we shall shortly revert, they are neither individually nor collectively compelling. Fortunately, there is also another sort of consideration that can lead scientists to reject an old paradigm in favor of a new. These are the arguments, rarely made entirely explicit, that appeal to the individual's sense of the appropriate or the aesthetic—the new theory is said to be "neater," "more suitable," or "simpler" than the old. Probably

[12] Kuhn, *op. cit.*, pp. 219–25.

[13] E. T. Whittaker, *A History of the Theories of Aether and Electricity*, I (2d ed.; London, 1951), 108.

[14] See *ibid.*, II (1953), 151–80, for the development of general relativity. For Einstein's reaction to the precise agreement of the theory with the observed motion of Mercury's perihelion, see the letter quoted in P. A. Schilpp (ed.), *Albert Einstein, Philosopher-Scientist* (Evanston, Ill., 1949), p. 101.

such arguments are less effective in the sciences than in mathematics. The early versions of most new paradigms are crude. By the time their full aesthetic appeal can be developed, most of the community has been persuaded by other means. Nevertheless, the importance of aesthetic considerations can sometimes be decisive. Though they often attract only a few scientists to a new theory, it is upon those few that its ultimate triumph may depend. If they had not quickly taken it up for highly individual reasons, the new candidate for paradigm might never have been sufficiently developed to attract the allegiance of the scientific community as a whole.

To see the reason for the importance of these more subjective and aesthetic considerations, remember what a paradigm debate is about. When a new candidate for paradigm is first proposed, it has seldom solved more than a few of the problems that confront it, and most of those solutions are still far from perfect. Until Kepler, the Copernican theory scarcely improved upon the predictions of planetary position made by Ptolemy. When Lavoisier saw oxygen as "the air itself entire," his new theory could cope not at all with the problems presented by the proliferation of new gases, a point that Priestley made with great success in his counterattack. Cases like Fresnel's white spot are extremely rare. Ordinarily, it is only much later, after the new paradigm has been developed, accepted, and exploited that apparently decisive arguments—the Foucault pendulum to demonstrate the rotation of the earth or the Fizeau experiment to show that light moves faster in air than in water—are developed. Producing them is part of normal science, and their role is not in paradigm debate but in postrevolutionary texts.

Before those texts are written, while the debate goes on, the situation is very different. Usually the opponents of a new paradigm can legitimately claim that even in the area of crisis it is little superior to its traditional rival. Of course, it handles some problems better, has disclosed some new regularities. But the older paradigm can presumably be articulated to meet these challenges as it has met others before. Both Tycho Brahe's earth-centered astronomical system and the later versions of the phlogiston theory were responses to challenges posed by a new candidate for paradigm, and both were quite successful.[15] In addition, the defenders of traditional theory and procedure can almost always point to problems that its new rival has not solved but that for their view are no problems at all. Until the discovery of the composition of water, the combustion of hydrogen was a strong argument for the phlogiston theory and against Lavoisier's. And after the oxygen theory had triumphed, it could still not explain the preparation of a combustible gas from carbon, a phenomenon to which the phlogistonists had pointed as strong support for their view.[16] Even in the area of crisis, the balance of argument and counterargument can sometimes be very close indeed. And outside that area the balance will often decisively favor the tradition. Copernicus destroyed a time-honored explanation of terrestrial motion without replacing it; Newton did the same for an older explanation of gravity, Lavoisier for the common properties of metals, and so on. In short, if a new candidate for paradigm had to be judged from the start by hardheaded people who examined only relative problem-solving ability, the sciences would experience very few major revolutions. Add the counterarguments generated by what we previously called the incommensurability of paradigms, and the sciences might experience no revolutions at all.

But paradigm debates are not really about relative problem-solving ability, though for good reasons they are usually couched in those terms. Instead, the issue is which paradigm should in the future guide research on problems many of which neither competitor can yet claim to resolve completely. A decision between alternate ways of practicing science is called for, and in the circumstances that decision must be based less on

[15] For Brahe's system, which was geometrically entirely equivalent to Copernicus', see J. L. E. Dreyer, *A History of Astronomy from Thales to Kepler* (2d ed.; New York, 1953), pp. 359–71. For the last versions of the phlogiston theory and their success, see J. R. Partington and D. McKie, "Historical Studies of the Phlogiston Theory," *Annals of Science*, IV (1939), 113–49.

[16] For the problem presented by hydrogen, see J. R. Partington, *A Short History of Chemistry* (2d ed.; London, 1951), p. 134. For carbon monoxide, see H. Kopp, *Geschichte der Chemie*, III (Braunschweig, 1845), 294–96.

past achievement than on future promise. The man who embraces a new paradigm at an early stage must often do so in defiance of the evidence provided by problem-solving. He must, that is, have faith that the new paradigm will succeed with the many large problems that confront it, knowing only that the older paradigm has failed with a few. A decision of that kind can only be made on faith.

That is one of the reasons why prior crisis proves so important. Scientists who have not experienced it will seldom renounce the hard evidence of problem-solving to follow what may easily prove and will be widely regarded as a will-o'-the-wisp. But crisis alone is not enough. There must also be a basis, though it need be neither rational nor ultimately correct, for faith in the particular candidate chosen. Something must make at least a few scientists feel that the new proposal is on the right track, and sometimes it is only personal and inarticulate aesthetic considerations that can do that. Men have been converted by them at times when most of the articulable technical arguments pointed the other way. When first introduced, neither Copernicus' astronomical theory nor De Broglie's theory of matter had many other significant grounds of appeal. Even today Einstein's general theory attracts men principally on aesthetic grounds, an appeal that few people outside of mathematics have been able to feel.

This is not to suggest that new paradigms triumph ultimately through some mystical aesthetic. On the contrary, very few men desert a tradition for these reasons alone. Often those who do turn out to have been misled. But if a paradigm is ever to triumph it must gain some first supporters, men who will develop it to the point where hardheaded arguments can be produced and multiplied. And even those arguments, when they come, are not individually decisive. Because scientists are reasonable men, one or another argument will ultimately persuade many of them. But there is no single argument that can or should persuade them all. Rather than a single group conversion, what occurs is an increasing shift in the distribution of professional allegiances.

At the start a new candidate for paradigm may have few supporters, and on occasions the supporters' motives may be suspect. Nevertheless, if they are competent, they will improve it, explore its possibilities, and show what it would be like to belong to the community guided by it. And as that goes on, if the paradigm is one destined to win its fight, the number and strength of the persuasive arguments in its favor will increase. More scientists will then be converted, and the exploration of the new paradigm will go on. Gradually the number of experiments, instruments, articles, and books based upon the paradigm will multiply. Still more men, convinced of the new view's fruitfulness, will adopt the new mode of practicing normal science, until at last only a few elderly hold-outs remain. And even they, we cannot say, are wrong. Though the historian can always find men—Priestley, for instance—who were unreasonable to resist for as long as they did, he will not find a point at which resistance becomes illogical or unscientific. At most he may wish to say that the man who continues to resist after his whole profession has been converted has *ipso facto* ceased to be a scientist.

XIII. Progress through Revolutions

The preceding pages have carried my schematic description of scientific development as far as it can go in this essay. Nevertheless, they cannot quite provide a conclusion. If this description has at all caught the essential structure of a science's continuing evolution, it will simultaneously have posed a special problem: Why should the enterprise sketched above move steadily ahead in ways that, say, art, political theory, or philosophy does not? Why is progress a perquisite reserved almost exclusively for the activities we call science? The most usual answers to that question have been denied in the body of this essay. We must conclude it by asking whether substitutes can be found.

Notice immediately that part of the question is entirely semantic. To a very great extent the term 'science' is reserved for fields that do progress in obvious ways. Nowhere does this show more clearly than in the recurrent debates about whether one or another of the contemporary social sciences is really a science. These debates have parallels in the pre-paradigm periods of fields that are today unhesitatingly labeled science. Their ostensible issue throughout is a definition of that vexing term. Men argue that psychology, for example, is a science because it possesses such and such characteristics. Others counter that those characteristics are either unnecessary or not sufficient to make a field a science. Often great energy is invested, great passion aroused, and the outsider is at a loss to know why. Can very much depend upon a *definition* of 'science'? Can a definition tell a man whether he is a scientist or not? If so, why do not natural scientists or artists worry about the definition of the term? Inevitably one suspects that the issue is more fundamental. Probably questions like the following are really being asked: Why does my field fail to move ahead in the way that, say, physics does? What changes in technique or method or ideology would enable it to do so? These are not, however, questions that could respond to an agreement on definition. Furthermore, if precedent from the natural sciences serves, they will cease to be a source of concern not when a definition is found, but when the groups that now doubt their own status achieve consensus about their past and present accomplishments. It may, for example, be significant that economists argue less about whether their field is a science than do practitioners of some other fields of social science. Is that because economists know what science is? Or is it rather economics about which they agree?

That point has a converse that, though no longer simply semantic, may help to display the inextricable connections between our notions of science and of progress. For many centuries, both in antiquity and again in early modern Europe, painting was regarded as *the* cumulative discipline. During those years the artist's goal was assumed to be representation. Critics and historians, like Pliny and Vasari, then recorded with veneration the series of inventions from foreshortening through chiaroscuro that had made possible successively more perfect representations of nature.[1] But those are also the years, particularly during the Renaissance, when little cleavage was felt between the sciences and the arts. Leonardo was only one of many men who passed freely back and forth between fields that only later became categorically distinct.[2] Furthermore, even after that steady exchange had ceased, the term 'art' continued to apply as much to technology and the crafts, which were also seen as progressive, as to painting and sculpture. Only when the latter unequivocally renounced representation as their goal and began to learn again from primitive models did the cleavage we now take for granted assume anything like its present depth. And even today, to switch fields once more, part of our difficulty in seeing the profound differences between science and technology must relate to the fact that progress is an obvious attribute of both fields.

[1] E. H. Gombrich, *Art and Illusion: A Study in the Psychology of Pictorial Representation* (New York, 1960), pp. 11–12.

[2] *Ibid.*, p. 97; and Giorgio de Santillana, "The Role of Art in the Scientific Renaissance," in *Critical Problems in the History of Science*, ed. M. Clagett (Madison, Wis., 1959), pp. 33–65.

It can, however, only clarify, not solve, our present difficulty to recognize that we tend to see as science any field in which progress is marked. There remains the problem of understanding why progress should be so noteworthy a characteristic of an enterprise conducted with the techniques and goals this essay has described. That question proves to be several in one, and we shall have to consider each of them separately. In all cases but the last, however, their resolution will depend in part upon an inversion of our normal view of the relation between scientific activity and the community that practices it. We must learn to recognize as causes what have ordinarily been taken to be effects. If we can do that, the phrases 'scientific progress' and even 'scientific objectivity' may come to seem in part redundant. In fact, one aspect of the redundancy has just been illustrated. Does a field make progress because it is a science, or is it a science because it makes progress?

Ask now why an enterprise like normal science should progress, and begin by recalling a few of its most salient characteristics. Normally, the members of a mature scientific community work from a single paradigm or from a closely related set. Very rarely do different scientific communities investigate the same problems. In those exceptional cases the groups hold several major paradigms in common. Viewed from within any single community, however, whether of scientists or of non-scientists, the result of successful creative work *is* progress. How could it possibly be anything else? We have, for example, just noted that while artists aimed at representation as their goal, both critics and historians chronicled the progress of the apparently united group. Other creative fields display progress of the same sort. The theologian who articulates dogma or the philosopher who refines the Kantian imperatives contributes to progress, if only to that of the group that shares his premises. No creative school recognizes a category of work that is, on the one hand, a creative success, but is not, on the other, an addition to the collective achievement of the group. If we doubt, as many do, that non-scientific fields make progress, that cannot be because individual schools make none. Rather, it must be because there are always competing schools, each of which constantly questions the very foundations of the others. The man who argues that philosophy, for example, has made no progress emphasizes that there are still Aristotelians, not that Aristotelianism has failed to progress.

These doubts about progress arise, however, in the sciences too. Throughout the pre-paradigm period when there is a multiplicity of competing schools, evidence of progress, except within schools, is very hard to find. This is the period described in Section II as one during which individuals practice science, but in which the results of their enterprise do not add up to science as we know it. And again, during periods of revolution when the fundamental tenets of a field are once more at issue, doubts are repeatedly expressed about the very possibility of continued progress if one or another of the opposed paradigms is adopted. Those who rejected Newtonianism proclaimed that its reliance upon innate forces would return science to the Dark Ages. Those who opposed Lavoisier's chemistry held that the rejection of chemical "principles" in favor of laboratory elements was the rejection of achieved chemical explanation by those who would take refuge in a mere name. A similar, though more moderately expressed, feeling seems to underlie the opposition of Einstein, Bohm, and others, to the dominant probabilistic interpretation of quantum mechanics. In short, it is only during periods of normal science that progress seems both obvious and assured. During those periods, however, the scientific community could view the fruits of its work in no other way.

With respect to normal science, then, part of the answer to the problem of progress lies simply in the eye of the beholder. Scientific progress is not different in kind from progress in other fields, but the absence at most times of competing schools that question each other's aims and standards makes the progress of a normal-scientific community far easier to see. That, however, is only part of the answer and by no means the most important part. We have, for example, already noted that once the reception of a common paradigm has freed the scientific community from the need constantly to re-examine its first principles, the members of that community can concentrate exclusively upon the subtlest and most esoteric of the phenomena that concern it. Inevitably, that does increase both the effectiveness and the efficiency with which the group as a whole solves new problems. Other aspects of professional life in the sciences enhance this very special efficiency still further.

Some of these are consequences of the unparalleled insulation of mature scientific communities from the demands of the laity and of everyday life. That insulation has never been complete—we are now discussing matters of degree. Nevertheless, there are no other professional communities in which individual creative work is so exclusively addressed to and evaluated by other members of the profession. The most esoteric of poets or the most abstract of theologians is far more concerned than the scientist with lay approbation of his creative work, though he may be even less concerned with approbation in general. That difference proves consequential. Just because he is working only for an audience of colleagues, an audience that shares his own values and beliefs, the scientist can take a single set of standards for granted. He need not worry about what some other group or school will think and can therefore dispose of one problem and get on to the next more quickly than those who work for a more heterodox group. Even more important, the insulation of the scientific community from society permits the individual scientist to concentrate his attention upon problems that he has good reason to believe he will be able to solve. Unlike the engineer, and many doctors, and most theologians, the scientist need not choose problems because they urgently need solution and without regard for the tools available to solve them. In this respect, also, the contrast between natural scientists and many social scientists proves instructive. The latter often tend, as the former almost never do, to defend their choice of a research problem—e.g., the effects of racial discrimination or the causes of the business cycle—chiefly in terms of the social importance of achieving a solution. Which group would one then expect to solve problems at a more rapid rate?

The effects of insulation from the larger society are greatly intensified by another characteristic of the professional scientific community, the nature of its educational initiation. In music, the graphic arts, and literature, the practitioner gains his education by exposure to the works of other artists, principally earlier artists. Textbooks, except compendia of or handbooks to original creations, have only a secondary role. In history, philosophy, and the social sciences, textbook literature has a greater significance. But even in these fields the elementary college course employs parallel readings in original sources, some of them the "classics" of the field, others the contemporary research reports that practitioners write for each other. As a result, the student in any one of these disciplines is constantly made aware of the immense variety of problems that the members of his future group have, in the course of time, attempted to solve. Even more important, he has constantly before him a number of competing and incommensurable solutions to these problems, solutions that he must ultimately evaluate for himself.

Contrast this situation with that in at least the contemporary natural sciences. In these fields the student relies mainly on textbooks until, in his third or fourth year of graduate work, he begins his own research. Many science curricula do not ask even graduate students to read in works not written specially for students. The few that do assign supplementary reading in research papers and monographs restrict such assignments to the most advanced courses and to materials that take up more or less where the available texts leave off. Until the very last stages in the education of a scientist, textbooks are systematically substituted for the creative scientific literature that made them possible. Given the confidence in their paradigms, which makes this educational technique possible, few scientists would wish to change it. Why, after all, should the student of physics, for example, read the works of Newton, Faraday, Einstein, or Schrödinger, when everything he needs to know about these works is recapitulated in a far briefer, more precise, and more systematic form in a number of up-to-date textbooks?

Without wishing to defend the excessive lengths to which this type of education has occasionally been carried, one cannot help but notice that in general it has been immensely effective.

Of course, it is a narrow and rigid education, probably more so than any other except perhaps in orthodox theology. But for normal-scientific work, for puzzle-solving within the tradition that the textbooks define, the scientist is almost perfectly equipped. Furthermore, he is well equipped for another task as well—the generation through normal science of significant crises. When they arise, the scientist is not, of course, equally well prepared. Even though prolonged crises are probably reflected in less rigid educational practice, scientific training is not well designed to produce the man who will easily discover a fresh approach. But so long as somebody appears with a new candidate for paradigm—usually a young man or one new to the field—the loss due to rigidity accrues only to the individual. Given a generation in which to effect the change, individual rigidity is compatible with a community that can switch from paradigm to paradigm when the occasion demands. Particularly, it is compatible when that very rigidity provides the community with a sensitive indicator that something has gone wrong.

In its normal state, then, a scientific community is an immensely efficient instrument for solving the problems or puzzles that its paradigms define. Furthermore, the result of solving those problems must inevitably be progress. There is no problem here. Seeing that much, however, only highlights the second main part of the problem of progress in the sciences. Let us therefore turn to it and ask about progress through extraordinary science. Why should progress also be the apparently universal concomitant of scientific revolutions? Once again, there is much to be learned by asking what else the result of a revolution could be. Revolutions close with a total victory for one of the two opposing camps. Will that group ever say that the result of its victory has been something less than progress? That would be rather like admitting that they had been wrong and their opponents right. To them, at least, the outcome of revolution must be progress, and they are in an excellent position to make certain that future members of their community will see past history in the same way. Section XI described in detail the techniques by which this is accomplished, and we have just recurred to a closely related aspect of professional scientific life. When it repudiates a past paradigm, a scientific community simultaneously renounces, as a fit subject for professional scrutiny, most of the books and articles in which that paradigm had been embodied. Scientific education makes use of no equivalent for the art museum or the library of classics, and the result is a sometimes drastic distortion in the scientist's perception of his discipline's past. More than the practitioners of other creative fields, he comes to see it as leading in a straight line to the discipline's present vantage. In short, he comes to see it as progress. No alternative is available to him while he remains in the field.

Inevitably those remarks will suggest that the member of a mature scientific community is, like the typical character of Orwell's *1984*, the victim of a history rewritten by the powers that be. Furthermore, that suggestion is not altogether inappropriate. There are losses as well as gains in scientific revolutions, and scientists tend to be peculiarly blind to the former.[3] On the other hand, no explanation of progress through revolutions may stop at this point. To do so would be to imply that in the sciences might makes right, a formulation which would again not be entirely wrong if it did not suppress the nature of the process and of the authority by which the choice between paradigms is made. If authority alone, and particularly if nonprofessional authority, were the arbiter of paradigm debates, the outcome of those debates might still be revolution, but it would not be *scientific* revolution. The very existence of science depends upon vesting the power to choose between paradigms in the members of a special kind of community. Just how special that community must be if science is to survive and grow may be indicated by the very tenuousness of humanity's hold on the scientific enterprise. Every civilization of which we have records

[3] Historians of science often encounter this blindness in a particularly striking form. The group of students who come to them from the sciences is very often the most rewarding group they teach. But it is also usually the most frustrating at the start. Because science students "know the right answers," it is particularly difficult to make them analyze an older science in its own terms.

has possessed a technology, an art, a religion, a political system, laws, and so on. In many cases those facets of civilization have been as developed as our own. But only the civilizations that descend from Hellenic Greece have possessed more than the most rudimentary science. The bulk of scientific knowledge is a product of Europe in the last four centuries. No other place and time has supported the very special communities from which scientific productivity comes.

What are the essential characteristics of these communities? Obviously, they need vastly more study. In this area only the most tentative generalizations are possible. Nevertheless, a number of requisites for membership in a professional scientific group must already be strikingly clear. The scientist must, for example, be concerned to solve problems about the behavior of nature. In addition, though his concern with nature may be global in its extent, the problems on which he works must be problems of detail. More important, the solutions that satisfy him may not be merely personal but must instead be accepted as solutions by many. The group that shares them may not, however, be drawn at random from society as a whole, but is rather the well-defined community of the scientist's professional compeers. One of the strongest, if still unwritten, rules of scientific life is the prohibition of appeals to heads of state or to the populace at large in matters scientific. Recognition of the existence of a uniquely competent professional group and acceptance of its role as the exclusive arbiter of professional achievement has further implications. The group's members, as individuals and by virtue of their shared training and experience, must be seen as the sole possessors of the rules of the game or of some equivalent basis for unequivocal judgments. To doubt that they shared some such basis for evaluations would be to admit the existence of incompatible standards of scientific achievement. That admission would inevitably raise the question whether truth in the sciences can be one.

This small list of characteristics common to scientific communities has been drawn entirely from the practice of normal science, and it should have been. That is the activity for which the scientist is ordinarily trained. Note, however, that despite its small size the list is already sufficient to set such communities apart from all other professional groups. And note, in addition, that despite its source in normal science the list accounts for many special features of the group's response during revolutions and particularly during paradigm debates. We have already observed that a group of this sort must see a paradigm change as progress. Now we may recognize that the perception is, in important respects, self-fulfilling. The scientific community is a supremely efficient instrument for maximizing the number and precision of the problem solved through paradigm change.

Because the unit of scientific achievement is the solved problem and because the group knows well which problems have already been solved, few scientists will easily be persuaded to adopt a viewpoint that again opens to question many problems that had previously been solved. Nature itself must first undermine professional security by making prior achievements seem problematic. Furthermore, even when that has occurred and a new candidate for paradigm has been evoked, scientists will be reluctant to embrace it unless convinced that two all-important conditions are being met. First, the new candidate must seem to resolve some outstanding and generally recognized problem that can be met in no other way. Second, the new paradigm must promise to preserve a relatively large part of the concrete problem-solving ability that has accrued to science through its predecessors. Novelty for its own sake is not a desideratum in the sciences as it is in so many other creative fields. As a result, though new paradigms seldom or never possess all the capabilities of their predecessors, they usually preserve a great deal of the most concrete parts of past achievement and they always permit additional concrete problem-solutions besides.

To say this much is not to suggest that the ability to solve problems is either the unique or an unequivocal basis for paradigm choice. We have already noted many reasons why there can be no criterion of that sort. But it does suggest that a community of scientific specialists will do all that it can to ensure the continuing growth of the assembled data that it can treat with precision and detail. In the process the community will sustain losses. Often some old problems must be banished. Frequently, in addition, revolution narrows the scope of the community's professional concerns, increases the extent of its specialization, and attenuates its communication with other groups, both scientific and lay. Though science surely grows in depth, it may not grow in breadth as well. If it does so, that breadth is manifest mainly in the proliferation of scientific specialties, not in the scope of any single specialty alone. Yet despite these and other losses to the individual communities, the nature of such communities provides a virtual guarantee that both the list of problems solved by science and the precision of individual problem-solutions will grow and grow. At least, the nature of the community provides such a guarantee if there is any way at all in which it can be provided. What better criterion than the decision of the scientific group could there be?

These last paragraphs point the directions in which I believe a more refined solution of the problem of progress in the sciences must be sought. Perhaps they indicate that scientific progress is not quite what we had taken it to be. But they simultaneously show that a sort of progress will inevitably characterize the scientific enterprise so long as such an enterprise survives. In the sciences there need not be progress of another sort. We may, to be more precise, have to relinquish the notion, explicit or implicit, that changes of paradigm carry scientists and those who learn from them closer and closer to the truth.

It is now time to notice that until the last very few pages the term 'truth' had entered this essay only in a quotation from Francis Bacon. And even in those pages it entered only as a source for the scientist's conviction that incompatible rules for doing science cannot coexist except during revolutions when the profession's main task is to eliminate all sets but one. The developmental process described in this essay has been a process of evolution *from* primitive beginnings—a process whose successive stages are characterized by an increasingly detailed and refined understanding of nature. But nothing that has been or will be said makes it a process of evolution *toward* anything. Inevitably that lacuna will have disturbed many readers. We are all deeply accustomed to seeing science as the one enterprise that draws constantly nearer to some goal set by nature in advance.

But need there be any such goal? Can we not account for both science's existence and its success in terms of evolution from the community's state of knowledge at any given time? Does it really help to imagine that there is some one full, objective, true account of nature and that the proper measure of scientific achievement is the extent to which it brings us closer to that ultimate goal? If we can learn to substitute evolution-from-what-we-do-know for evolution-toward-what-we-wish-to-know, a number of vexing problems may vanish in the process. Somewhere in this maze, for example, must lie the problem of induction.

I cannot yet specify in any detail the consequences of this alternate view of scientific advance. But it helps to recognize that the conceptual transposition here recommended is very close to one that the West undertook just a century ago. It is particularly helpful because in both cases the main obstacle to transposition is the same. When Darwin first published his theory of evolution by natural selection in 1859, what most bothered many professionals was neither the notion of species change nor the possible descent of man from apes. The evidence pointing to evolution, including the evolution of man, had been accumulating for decades, and the idea of evolution had been suggested and widely disseminated before. Though evolution, as such, did encounter resistance, particularly from some religious groups, it was by no means the greatest of the difficulties the Darwinians faced. That difficulty stemmed from an idea that was more nearly Darwin's own. All the well-known pre-Darwinian evolutionary theories—those of Lamarck, Chambers, Spencer, and the German *Naturphilosophen*—had taken evolution to be a goal-directed process. The "idea" of man and of the contemporary flora and fauna was thought to have been present from the first creation of life, perhaps in the mind of God. That idea or plan had provided the direction and the guiding force to the entire evolutionary process. Each new stage of evolutionary development was a more perfect realization of a plan that had been present from the start.[4]

For many men the abolition of that teleological kind of evolution was the most significant and least palatable of Darwin's suggestions.[5] The *Origin of Species* recognized no goal set either by God or nature. Instead, natural selection, operating in the given environment and with the actual organisms presently at hand, was responsible for the gradual but steady emergence of more elaborate, further articulated, and vastly more specialized organisms. Even such marvelously adapted organs as the eye and hand of man—organs whose design had previously provided powerful arguments for the existence of a supreme artificer and an advance plan—were products of a process that moved steadily *from* primitive beginnings but *toward* no goal. The belief that natural selection, resulting from mere competition between organisms for survival, could have produced man together with the higher animals and plants was the most difficult and disturbing aspect of Darwin's theory. What could 'evolution,' 'development,' and 'progress' mean in the absence of a specified goal? To many people, such terms suddenly seemed self-contradictory.

The analogy that relates the evolution of organisms to the evolution of scientific ideas can easily be pushed too far. But with respect to the issues of this closing section it is very nearly perfect. The process described in Section XII as the resolution of revolutions is the selection by conflict within the scientific community of the fittest way to practice future science. The net result of a sequence of such revolutionary selections, separated by periods of normal research, is the wonderfully adapted set of instruments we call modern scientific knowledge. Successive stages in that developmental process are marked by an increase in articulation and specialization. And the entire process may have occurred, as we now suppose biological evolution did,

[4] Loren Eiseley, *Darwin's Century: Evolution and the Men Who Discovered It* (New York, 1958), chaps. ii, iv–v.

[5] For a particularly acute account of one prominent Darwinian's struggle with this problem, see A. Hunter Dupree, *Asa Gray, 1810–1888* (Cambridge, Mass., 1959), pp. 295–306, 355–83.

without benefit of a set goal, a permanent fixed scientific truth, of which each stage in the development of scientific knowledge is a better exemplar.

Anyone who has followed the argument this far will nevertheless feel the need to ask why the evolutionary process should work. What must nature, including man, be like in order that science be possible at all? Why should scientific communities be able to reach a firm consensus unattainable in other fields? Why should consensus endure across one paradigm change after another? And why should paradigm change invariably produce an instrument more perfect in any sense than those known before? From one point of view those questions, excepting the first, have already been answered. But from another they are as open as they were when this essay began. It is not only the scientific community that must be special. The world of which that community is a part must also possess quite special characteristics, and we are no closer than we were at the start to knowing what these must be. That problem—What must the world be like in order that man may know it?—was not, however, created by this essay. On the contrary, it is as old as science itself, and it remains unanswered. But it need not be answered in this place. Any conception of nature compatible with the growth of science by proof is compatible with the evolutionary view of science developed here. Since this view is also compatible with close observation of scientific life, there are strong arguments for employing it in attempts to solve the host of problems that still remain.

Postscript—1969

It has now been almost seven years since this book was first published.[1] In the interim both the response of critics and my own further work have increased my understanding of a number of the issues it raises. On fundamentals my viewpoint is very nearly unchanged, but I now recognize aspects of its initial formulation that create gratuitous difficulties and misunderstandings. Since some of those misunderstandings have been my own, their elimination enables me to gain ground that should ultimately provide the basis for a new version of the book.[2] Meanwhile, I welcome the chance to sketch needed revisions, to comment on some reiterated criticisms, and to suggest directions in which my own thought is presently developing.[3]

Several of the key difficulties of my original text cluster about the concept of a paradigm, and my discussion begins with them.[4] In the subsection that follows at once, I suggest the desirability of disentangling that concept from the notion of a scientific community, indicate how this may be done, and discuss some signifi-

[1] This postscript was first prepared at the suggestion of my onetime student and longtime friend, Dr. Shigeru Nakayama of the University of Tokyo, for inclusion in his Japanese translation of this book. I am grateful to him for the idea, for his patience in awaiting its fruition, and for permission to include the result in the English language edition.

[2] For this edition I have attempted no systematic rewriting, restricting alterations to a few typographical errors plus two passages which contained isolable errors. One of these is the description of the role of Newton's *Principia* in the development of eighteenth-century mechanics on pp. 30–33, above. The other concerns the response to crises on p. 84.

[3] Other indications will be found in two recent essays of mine: "Reflection on My Critics," in Imre Lakatos and Alan Musgrave (eds.), *Criticism and the Growth of Knowledge* (Cambridge, 1970); and "Second Thoughts on Paradigms," in Frederick Suppe (ed.), *The Structure of Scientific Theories* (Urbana, Ill., 1970 or 1971), both currently in press. I shall cite the first of these essays below as "Reflections" and the volume in which it appears as *Growth of Knowledge;* the second essay will be referred to as "Second Thoughts."

[4] For particularly cogent criticism of my initial presentation of paradigms see: Margaret Masterman, "The Nature of a Paradigm," in *Growth of Knowledge;* and Dudley Shapere, "The Structure of Scientific Revolutions," *Philosophical Review,* LXXIII (1964), 383–94.

cant consequences of the resulting analytic separation. Next I consider what occurs when paradigms are sought by examining the behavior of the members of a *previously determined* scientific community. That procedure quickly discloses that in much of the book the term 'paradigm' is used in two different senses. On the one hand, it stands for the entire constellation of beliefs, values, techniques, and so on shared by the members of a given community. On the other, it denotes one sort of element in that constellation, the concrete puzzle-solutions which, employed as models or examples, can replace explicit rules as a basis for the solution of the remaining puzzles of normal science. The first sense of the term, call it the sociological, is the subject of Subsection 2, below; Subsection 3 is devoted to paradigms as exemplary past achievements.

Philosophically, at least, this second sense of 'paradigm' is the deeper of the two, and the claims I have made in its name are the main sources for the controversies and misunderstandings that the book has evoked, particularly for the charge that I make of science a subjective and irrational enterprise. These issues are considered in Subsections 4 and 5. The first argues that terms like 'subjective' and 'intuitive' cannot appropriately be applied to the components of knowledge that I have described as tacitly embedded in shared examples. Though such knowledge is not, without essential change, subject to paraphrase in terms of rules and criteria, it is nevertheless systematic, time tested, and in some sense corrigible. Subsection 5 applies that argument to the problem of choice between two incompatible theories, urging in brief conclusion that men who hold incommensurable viewpoints be thought of as members of different language communities and that their communication problems be analyzed as problems of translation. Three residual issues are discussed in the concluding Subsections, 6 and 7. The first considers the charge that the view of science developed in this book is through-and-through relativistic. The second begins by inquiring whether my argument really suffers, as has been said, from a confusion between the descriptive and the normative modes; it concludes with brief remarks on a topic deserving a separate essay: the extent to which the book's main theses may legitimately be applied to fields other than science.

1. *Paradigms and Community Structure*

The term 'paradigm' enters the preceding pages early, and its manner of entry is intrinsically circular. A paradigm is what the members of a scientific community share, *and*, conversely, a scientific community consists of men who share a paradigm. Not all circularities are vicious (I shall defend an argument of similar structure late in this postscript), but this one is a source of real difficulties. Scientific communities can and should be isolated without prior recourse to paradigms; the latter can then be discovered by scrutinizing the behavior of a given community's members. If this book were being rewritten, it would therefore open with a discussion of the community structure of science, a topic that has recently become a significant subject of sociological research and that historians of science are also beginning to take seriously. Preliminary results, many of them still unpublished, suggest that the empirical techniques required for its exploration are non-trivial, but some are in hand and others are sure to be developed.[5] Most practicing scientists respond at once to questions about their community affiliations, taking for granted that responsibility for the various current specialties is distributed among groups of at least roughly determinate membership. I shall therefore here assume that more systematic means for their identification will be found. Instead of presenting preliminary research results, let me briefly articulate the intuitive notion of community that underlies much in the earlier chapters of this book. It is a notion now widely shared by scientists, sociologists, and a number of historians of science.

[5] W. O. Hagstrom, *The Scientific Community* (New York, 1965), chaps. iv and v; D. J. Price and D. de B. Beaver, "Collaboration in an Invisible College," *American Psychologist,* XXI (1966), 1011–18; Diana Crane, "Social Structure in a Group of Scientists: A Test of the 'Invisible College' Hypothesis," *American Sociological Review,* XXXIV (1969), 335–52; N. C. Mullins, *Social Networks among Biological Scientists,* (Ph.D. diss., Harvard University, 1966), and "The Micro-Structure of an Invisible College: The Phage Group" (paper delivered at an annual meeting of the American Sociological Association, Boston, 1968).

A scientific community consists, on this view, of the practitioners of a scientific specialty. To an extent unparalleled in most other fields, they have undergone similar educations and professional initiations; in the process they have absorbed the same technical literature and drawn many of the same lessons from it. Usually the boundaries of that standard literature mark the limits of a scientific subject matter, and each community ordinarily has a subject matter of its own. There are schools in the sciences, communities, that is, which approach the same subject from incompatible viewpoints. But they are far rarer there than in other fields; they are always in competition; and their competition is usually quickly ended. As a result, the members of a scientific community see themselves and are seen by others as the men uniquely responsible for the pursuit of a set of shared goals, including the training of their successors. Within such groups communication is relatively full and professional judgment relatively unanimous. Because the attention of different scientific communities is, on the other hand, focused on different matters, professional communication across group lines is sometimes arduous, often results in misunderstanding, and may, if pursued, evoke significant and previously unsuspected disagreement.

Communities in this sense exist, of course, at numerous levels. The most global is the community of all natural scientists. At an only slightly lower level the main scientific professional groups are communities: physicists, chemists, astronomers, zoologists, and the like. For these major groupings, community membership is readily established except at the fringes. Subject of highest degree, membership in professional societies, and journals read are ordinarily more than sufficient. Similar techniques will also isolate major subgroups: organic chemists, and perhaps protein chemists among them, solid-state and high-energy physicists, radio astronomers, and so on. It is only at the next lower level that empirical problems emerge. How, to take a contemporary example, would one have isolated the phage group prior to its public acclaim? For this purpose one must have recourse to attendance at special conferences, to the distribution of draft manuscripts or galley proofs prior to publication, and above all to formal and informal communication networks including those discovered in correspondence and in the linkages among citations.[6] I take it that the job can and will be done, at least for the contemporary scene and the more recent parts of the historical. Typically it may yield communities of perhaps one hundred members, occasionally significantly fewer. Usually individual scientists, particularly the ablest, will belong to several such groups either simultaneously or in succession.

Communities of this sort are the units that this book has presented as the producers and validators of scientific knowledge. Paradigms are something shared by the members of such groups. Without reference to the nature of these shared elements, many aspects of science described in the preceding pages can scarcely be understood. But other aspects can, though they are not independently presented in my original text. It is therefore worth noting, before turning to paradigms directly, a series of issues that require reference to community structure alone.

Probably the most striking of these is what I have previously called the transition from the pre- to the post-paradigm period in the development of a scientific field. That transition is the one sketched above in Section II. Before it occurs, a number of schools compete for the domination of a given field. Afterward, in the wake of some notable scientific achievement, the number of schools is greatly reduced, ordinarily to one, and a more efficient mode of scientific practice begins. The latter is generally esoteric and oriented to puzzle-solving, as the work of a group can be only when its members take the foundations of their field for granted.

The nature of that transition to maturity deserves fuller discussion than it has received in this book, particularly from those concerned with the development of the contemporary social

[6] Eugene Garfield, *The Use of Citation Data in Writing the History of Science* (Philadelphia: Institute of Scientific Information, 1964); M. M. Kessler, "Comparison of the Results of Bibliographic Coupling and Analytic Subject Indexing," *American Documentation*, XVI (1965), 223–33; D. J. Price, "Networks of Scientific Papers," *Science*, CIL (1965), 510–15.

sciences. To that end it may help to point out that the transition need not (I now think should not) be associated with the first acquisition of a paradigm. The members of all scientific communities, including the schools of the "pre-paradigm" period, share the sorts of elements which I have collectively labelled 'a paradigm.' What changes with the transition to maturity is not the presence of a paradigm but rather its nature. Only after the change is normal puzzle-solving research possible. Many of the attributes of a developed science which I have above associated with the acquisition of a paradigm I would therefore now discuss as consequences of the acquisition of the sort of paradigm that identifies challenging puzzles, supplies clues to their solution, and guarantees that the truly clever practitioner will succeed. Only those who have taken courage from observing that their own field (or school) has paradigms are likely to feel that something important is sacrificed by the change.

A second issue, more important at least to historians, concerns this book's implicit one-to-one identification of scientific communities with scientific subject matters. I have, that is, repeatedly acted as though, say, 'physical optics,' 'electricity,' and 'heat' must name scientific communities because they do name subject matters for research. The only alternative my text has seemed to allow is that all these subjects have belonged to the physics community. Identifications of that sort will not, however, usually withstand examination, as my colleagues in history have repeatedly pointed out. There was, for example, no physics community before the mid-nineteenth century, and it was then formed by the merger of parts of two previously separate communities, mathematics and natural philosophy *(physique expérimentale)*. What is today the subject matter for a single broad community has been variously distributed among diverse communities in the past. Other narrower subjects, for example heat and the theory of matter, have existed for long periods without becoming the special province of any single scientific community. Both normal science and revolutions are, however, community-based activities. To discover and analyze them, one must first unravel the changing community structure of the sciences over time. A paradigm governs, in the first instance, not a subject matter but rather a group of practitioners. Any study of paradigm-directed or of paradigm-shattering research must begin by locating the responsible group or groups.

When the analysis of scientific development is approached in that way, several difficulties which have been foci for critical attention are likely to vanish. A number of commentators have, for example, used the theory of matter to suggest that I drastically overstate the unanimity of scientists in their allegiance to a paradigm. Until comparatively recently, they point out, those theories have been topics for continuing disagreement and debate. I agree with the description but think it no counterexample. Theories of matter were not, at least until about 1920, the special province or the subject matter for any scientific community. Instead, they were tools for a large number of specialists' groups. Members of different communities sometimes chose different tools and criticized the choice made by others. Even more important, a theory of matter is not the sort of topic on which the members of even a single community must necessarily agree. The need for agreement depends on what it is the community does. Chemistry in the first half of the nineteenth century provides a case in point. Though several of the community's fundamental tools—constant proportion, multiple proportion, and combining weights—had become common property as a result of Dalton's atomic theory, it was quite possible for chemists, after the event, to base their work on these tools and to disagree, sometimes vehemently, about the existence of atoms.

Some other difficulties and misunderstandings will, I believe, be dissolved in the same way. Partly because of the examples I have chosen and partly because of my vagueness about the nature and size of the relevant communities, a few readers of this book have concluded that my concern is primarily or exclusively with major revolutions such as those associated with Copernicus, Newton, Darwin, or Einstein. A clearer delineation of community structure should, however, help to enforce the rather different impression I have tried to create. A revolution is for me a special sort of change involving a certain sort of reconstruction of group commitments. But it need not be a large change, nor need it seem revolutionary to those outside a single community, consisting perhaps of fewer than twenty-five people. It is just because this type of change, little recognized or discussed in the literature of the philosophy of science, occurs so regularly on this smaller scale that revolutionary, as against cumulative, change so badly needs to be understood.

One last alteration, closely related to the preceding, may help to facilitate that understanding. A number of critics have doubted whether crisis, the common awareness that something has gone wrong, precedes revolutions so invariably as I have implied in my original text. Nothing important to my argument depends, however, on crises' being an absolute prerequisite to revolutions; they need only be the usual prelude, supplying, that is, a self-correcting mechanism which ensures that the rigidity of normal science will not forever go unchallenged. Revolutions may also be induced in other ways, though I think they seldom are. In addition, I would now point out what the absence of an adequate discussion of community structure has obscured above: crises need not be generated by the work of the community that experiences them and that sometimes undergoes revolution as a result. New instruments like the electron microscope or new laws like Maxwell's may develop in one specialty and their assimilation create crisis in another.

2. *Paradigms as the Constellation of Group Commitments*

Turn now to paradigms and ask what they can possibly be. My original text leaves no more obscure or important question. One sympathetic reader, who shares my conviction that 'paradigm' names the central philosophical elements of the book, prepared a partial analytic index and concluded that the term is used in at least twenty-two different ways.[7] Most of those differences are, I now think, due to stylistic inconsistencies (e.g., Newton's Laws are sometimes a paradigm, sometimes parts of a paradigm, and

[7] Masterman, *op. cit.*

sometimes paradigmatic), and they can be eliminated with relative ease. But, with that editorial work done, two very different usages of the term would remain, and they require separation. The more global use is the subject of this subsection; the other will be considered in the next.

Having isolated a particular community of specialists by techniques like those just discussed, one may usefully ask: What do its members share that accounts for the relative fulness of their professional communication and the relative unanimity of their professional judgments? To that question my original text licenses the answer, a paradigm or set of paradigms. But for this use, unlike the one to be discussed below, the term is inappropriate. Scientists themselves would say they share a theory or set of theories, and I shall be glad if the term can ultimately be recaptured for this use. As currently used in philosophy of science, however, 'theory' connotes a structure far more limited in nature and scope than the one required here. Until the term can be freed from its current implications, it will avoid confusion to adopt another. For present purposes I suggest 'disciplinary matrix': 'disciplinary' because it refers to the common possession of the practitioners of a particular discipline; 'matrix' because it is composed of ordered elements of various sorts, each requiring further specification. All or most of the objects of group commitment that my original text makes paradigms, parts of paradigms, or paradigmatic are constituents of the disciplinary matrix, and as such they form a whole and function together. They are, however, no longer to be discussed as though they were all of a piece. I shall not here attempt an exhaustive list, but noting the main sorts of components of a disciplinary matrix will both clarify the nature of my present approach and simultaneously prepare for my next main point.

One important sort of component I shall label 'symbolic generalizations,' having in mind those expressions, deployed without question or dissent by group members, which can readily be cast in a logical form like $(x)(y)(z)\phi(x, y, z)$. They are the formal or the readily formalizable components of the disciplinary matrix. Sometimes they are found already in symbolic form: $f = ma$ or $I = V/R$. Others are ordinarily expressed in words: "elements combine in constant proportion by weight," or "action equals reaction." If it were not for the general acceptance of expressions like these, there would be no points at which group members could attach the powerful techniques of logical and mathematical manipulation in their puzzle-solving enterprise. Though the example of taxonomy suggests that normal science can proceed with few such expressions, the power of a science seems quite generally to increase with the number of symbolic generalizations its practioners have at their disposal.

These generalizations look like laws of nature, but their function for group members is not often that alone. Sometimes it is: for example the Joule-Lenz Law, $H = RI^2$. When that law was discovered, community members already knew what H, R, and I stood for, and these generalizations simply told them something about the behavior of heat, current, and resistance that they had not known before. But more often, as discussion earlier in the book indicates, symbolic generalizations simultaneously serve a second function, one that is ordinarily sharply separated in analyses by philosophers of science. Like $f = ma$ or $I = V/R$, they function in part as laws but also in part as definitions of some of the symbols they deploy. Furthermore, the balance between their inseparable legislative and definitional force shifts over time. In another context these points would repay detailed analysis, for the nature of the commitment to a law is very different from that of commitment to a definition. Laws are often corrigible piecemeal, but definitions, being tautologies, are not. For example, part of what the acceptance of Ohm's Law demanded was a redefinition of both 'current' and 'resistance'; if those terms had continued to mean what they had meant before, Ohm's Law could not have been right; that is why it was so strenuously opposed as, say, the Joule-Lenz Law was not.[8] Probably that situation is typical. I currently suspect that

[8] For significant parts of this episode see: T. M. Brown, "The Electric Current in Early Nineteenth-Century French Physics," *Historical Studies in the Physical Sciences*, I (1969), 61–103, and Morton Schagrin, "Resistance to Ohm's Law," *American Journal of Physics*, XXI (1963), 536–47.

all revolutions involve, among other things, the abandonment of generalizations the force of which had previously been in some part that of tautologies. Did Einstein show that simultaneity was relative or did he alter the notion of simultaneity itself? Were those who heard paradox in the phrase 'relativity of simultaneity' simply wrong?

Consider next a second type of component of the disciplinary matrix, one about which a good deal has been said in my original text under such rubrics as 'metaphysical paradigms' or 'the metaphysical parts of paradigms.' I have in mind shared commitments to such beliefs as: heat is the kinetic energy of the constituent parts of bodies; all perceptible phenomena are due to the interaction of qualitatively neutral atoms in the void, or, alternatively, to matter and force, or to fields. Rewriting the book now I would describe such commitments as beliefs in particular models, and I would expand the category models to include also the relatively heuristic variety: the electric circuit may be regarded as a steady-state hydrodynamic system; the molecules of a gas behave like tiny elastic billiard balls in random motion. Though the strength of group commitment varies, with nontrivial consequences, along the spectrum from heuristic to ontological models, all models have similar functions. Among other things they supply the group with preferred or permissible analogies and metaphors. By doing so they help to determine what will be accepted as an explanation and as a puzzle-solution; conversely, they assist in the determination of the roster of unsolved puzzles and in the evaluation of the importance of each. Note, however, that the members of scientific communities may not have to share even heuristic models, though they usually do so. I have already pointed out that membership in the community of chemists during the first half of the nineteenth century did not demand a belief in atoms.

A third sort of element in the disciplinary matrix I shall here describe as values. Usually they are more widely shared among different communities than either symbolic generalizations or models, and they do much to provide a sense of community to natural scientists as a whole. Though they function at all times, their particular importance emerges when the members of a particular community must identify crisis or, later, choose between incompatible ways of practicing their discipline. Probably the most deeply held values concern predictions: they should be accurate; quantitative predictions are preferable to qualitative ones; whatever the margin of permissible error, it should be consistently satisfied in a given field; and so on. There are also, however, values to be used in judging whole theories: they must, first and foremost, permit puzzle-formulation and solution; where possible they should be simple, self-consistent, and plausible, compatible, that is, with other theories currently deployed. (I now think it a weakness of my original text that so little attention is given to such values as internal and external consistency in considering sources of crisis and factors in theory choice.) Other sorts of values exist as well—for example, science should (or need not) be socially useful—but the preceding should indicate what I have in mind.

One aspect of shared values does, however, require particular mention. To a greater extent than other sorts of components of the disciplinary matrix, values may be shared by men who differ in their application. Judgments of accuracy are relatively, though not entirely, stable from one time to another and from one member to another in a particular group. But judgments of simplicity, consistency, plausibility, and so on often vary greatly from individual to individual. What was for Einstein an insupportable inconsistency in the old quantum theory, one that rendered the pursuit of normal science impossible, was for Bohr and others a difficulty that could be expected to work itself out by normal means. Even more important, in those situations where values must be applied, different values, taken alone, would often dictate different choices. One theory may be more accurate but less consistent or plausible than another; again the old quantum theory provides an example. In short, though values are widely shared by scientists and though commitment to them is both deep and constitutive of science, the application of values is sometimes considerably affected by the features of individual personality and biography that differentiate the members of the group.

To many readers of the preceding chapters, this characteristic of the operation of shared values has seemed a major weakness of my position. Because I insist that what scientists share is not sufficient to command uniform assent about such matters as the choice between competing theories or the distinction between an ordinary anomaly and a crisis-provoking one, I am occasionally accused of glorifying subjectivity and even irrationality.[9] But that reaction ignores two characteristics displayed by value judgments in any field. First, shared values can be important determinants of group behavior even though the members of the group do not all apply them in the same way. (If that were not the case, there would be no *special* philosophic problems about value theory or aesthetics.) Men did not all paint alike during the periods when representation was a primary value, but the developmental pattern of the plastic arts changed drastically when that value was abandoned.[10] Imagine what would happen in the sciences if consistency ceased to be a primary value. Second, individual variability in the application of shared values may serve functions essential to science. The points at which values must be applied are invariably also those at which risks must be taken. Most anomalies are resolved by normal means; most proposals for new theories do prove to be wrong. If all members of a community responded to each anomaly as a source of crisis or embraced each new theory advanced by a colleague, science would cease. If, on the other hand, no one reacted to anomalies or to brand-new theories in high-risk ways, there would be few or no revolutions. In matters like these the resort to shared values rather than to shared rules governing individual choice may be the community's way of distributing risk and assuring the long-term success of its enterprise.

Turn now to a fourth sort of element in the disciplinary matrix, not the only other kind but the last I shall discuss here. For it the term 'paradigm' would be entirely appropriate, both philologi-

[9] See particularly: Dudley Shapere, "Meaning and Scientific Change," in *Mind and Cosmos: Essays in Contemporary Science and Philosophy*, The University of Pittsburgh Series in the Philosophy of Science, III (Pittsburgh, 1966), 41–85; Israel Scheffler, *Science and Subjectivity* (New York, 1967); and the essays of Sir Karl Popper and Imre Lakatos in *Growth of Knowledge*.

[10] See the discussion at the beginning of Section XIII, above.

cally and autobiographically; this is the component of a group's shared commitments which first led me to the choice of that word. Because the term has assumed a life of its own, however, I shall here substitute 'exemplars.' By it I mean, initially, the concrete problem-solutions that students encounter from the start of their scientific education, whether in laboratories, on examinations, or at the ends of chapters in science texts. To these shared examples should, however, be added at least some of the technical problem-solutions found in the periodical literature that scientists encounter during their post-educational research careers and that also show them by example how their job is to be done. More than other sorts of components of the disciplinary matrix, differences between sets of exemplars provide the community fine-structure of science. All physicists, for example, begin by learning the same exemplars: problems such as the inclined plane, the conical pendulum, and Keplerian orbits; instruments such as the vernier, the calorimeter, and the Wheatstone bridge. As their training develops, however, the symbolic generalizations they share are increasingly illustrated by different exemplars. Though both solid-state and field-theoretic physicists share the Schrödinger equation, only its more elementary applications are common to both groups.

3. *Paradigms as Shared Examples*

The paradigm as shared example is the central element of what I now take to be the most novel and least understood aspect of this book. Exemplars will therefore require more attention than the other sorts of components of the disciplinary matrix. Philosophers of science have not ordinarily discussed the problems encountered by a student in laboratories or in science texts, for these are thought to supply only practice in the application of what the student already knows. He cannot, it is said, solve problems at all unless he has first learned the theory and some rules for applying it. Scientific knowledge is embedded in theory and rules; problems are supplied to gain facility in their application. I have tried to argue, however, that this localization of the cognitive content of science is wrong. After the student has done many problems, he may gain only added facility by solving more. But at the start and for some time after, doing problems is learning consequential things about nature. In the absence of such exemplars, the laws and theories he has previously learned would have little empirical content.

To indicate what I have in mind I revert briefly to symbolic generalizations. One widely shared example is Newton's Second Law of Motion, generally written as $f = ma$. The sociologist, say, or the linguist who discovers that the corresponding expression is unproblematically uttered and received by the members of a given community will not, without much additional investigation, have learned a great deal about what either the expression or the terms in it mean, about how the scientists of the community attach the expression to nature. Indeed, the fact that they accept it without question and use it as a point at which to introduce logical and mathematical manipulation does not of itself imply that they agree at all about such matters as meaning and application. Of course they do agree to a considerable extent, or the fact would rapidly emerge from their subsequent conversation. But one may well ask at what point and by what means they have come to do so. How have they learned, faced with a given experimental situation, to pick out the relevant forces, masses, and accelerations?

In practice, though this aspect of the situation is seldom or never noted, what students have to learn is even more complex than that. It is not quite the case that logical and mathematical manipulation are applied directly to $f = ma$. That expression proves on examination to be a law-sketch or a law-schema. As the student or the practicing scientist moves from one problem situation to the next, the symbolic generalization to which such manipulations apply changes. For the case of free fall, $f = ma$ becomes $mg = m\frac{d^2s}{dt^2}$; for the simple pendulum it is transformed to $mg \sin\theta = -ml\frac{d^2\theta}{dt^2}$; for a pair of interacting harmonic oscillators it becomes two equations, the first of which may be written

$m_1 \frac{d^2s_1}{dt^2} + k_1s_1 = k_2(s_2 - s_1 + d)$; and for more complex situations, such as the gyroscope, it takes still other forms, the family resemblance of which to $f = ma$ is still harder to discover. Yet, while learning to identify forces, masses, and accelerations in a variety of physical situations not previously encountered, the student has also learned to design the appropriate version of $f = ma$ through which to interrelate them, often a version for which he has encountered no literal equivalent before. How has he learned to do this?

A phenomenon familiar to both students of science and historians of science provides a clue. The former regularly report that they have read through a chapter of their text, understood it perfectly, but nonetheless had difficulty solving a number of the problems at the chapter's end. Ordinarily, also, those difficulties dissolve in the same way. The student discovers, with or without the assistance of his instructor, a way to see his problem as *like* a problem he has already encountered. Having seen the resemblance, grasped the analogy between two or more distinct problems, he can interrelate symbols and attach them to nature in the ways that have proved effective before. The law-sketch, say $f = ma$, has functioned as a tool, informing the student what similarities to look for, signaling the gestalt in which the situation is to be seen. The resultant ability to see a variety of situations as like each other, as subject for $f = ma$ or some other symbolic generalization, is, I think, the main thing a student acquires by doing exemplary problems, whether with a pencil and paper or in a well-designed laboratory. After he has completed a certain number, which may vary widely from one individual to the next, he views the situations that confront him as a scientist in the same gestalt as other members of his specialists' group. For him they are no longer the same situations he had encountered when his training began. He has meanwhile assimilated a time-tested and group-licensed way of seeing.

The role of acquired similarity relations also shows clearly in the history of science. Scientists solve puzzles by modeling them on previous puzzle-solutions, often with only minimal recourse to symbolic generalizations. Galileo found that a ball rolling down an incline acquires just enough velocity to return it to the same vertical height on a second incline of any slope, and he learned to see that experimental situation as like the pendulum with a point-mass for a bob. Huyghens then solved the problem of the center of oscillation of a physical pendulum by imagining that the extended body of the latter was composed of Galilean point-pendula, the bonds between which could be instantaneously released at any point in the swing. After the bonds were released, the individual point-pendula would swing freely, but their collective center of gravity when each attained its highest point would, like that of Galileo's pendulum, rise only to the height from which the center of gravity of the extended pendulum had begun to fall. Finally, Daniel Bernoulli discovered how to make the flow of water from an orifice resemble Huyghens' pendulum. Determine the descent of the center of gravity of the water in tank and jet during an infinitesimal interval of time. Next imagine that each particle of water afterward moves separately upward to the maximum height attainable with the velocity acquired during that interval. The ascent of the center of gravity of the individual particles must then equal the descent of the center of gravity of the water in tank and jet. From that view of the problem the long-sought speed of efflux followed at once.[11]

That example should begin to make clear what I mean by learning from problems to see situations as like each other, as subjects for the application of the same scientific law or law-sketch. Simultaneously it should show why I refer to the consequential knowledge of nature acquired while learning the similarity relationship and thereafter embodied in a way of viewing

[11] For the example, see: René Dugas, *A History of Mechanics,* trans. J. R. Maddox (Neuchatel, 1955), pp. 135–36, 186–93, and Daniel Bernoulli, *Hydrodynamica, sive de viribus et motibus fluidorum, commentarii opus academicum* (Strasbourg, 1738), Sec. iii. For the extent to which mechanics progressed during the first half of the eighteenth century by modelling one problem-solution on another, see Clifford Truesdell, "Reactions of Late Baroque Mechanics to Success, Conjecture, Error, and Failure in Newton's *Principia,*" *Texas Quarterly,* X (1967), 238–58.

physical situations rather than in rules or laws. The three problems in the example, all of them exemplars for eighteenth-century mechanicians, deploy only one law of nature. Known as the Principle of *vis viva,* it was usually stated as: "Actual descent equals potential ascent." Bernoulli's application of the law should suggest how consequential it was. Yet the verbal statement of the law, taken by itself, is virtually impotent. Present it to a contemporary student of physics, who knows the words and can do all these problems but now employs different means. Then imagine what the words, though all well known, can have said to a man who did not know even the problems. For him the generalization could begin to function only when he learned to recognize "actual descents" and "potential ascents" as ingredients of nature, and that is to learn something, prior to the law, about the situations that nature does and does not present. That sort of learning is not acquired by exclusively verbal means. Rather it comes as one is given words together with concrete examples of how they function in use; nature and words are learned together. To borrow once more Michael Polanyi's useful phrase, what results from this process is "tacit knowledge" which is learned by doing science rather than by acquiring rules for doing it.

4. *Tacit Knowledge and Intuition*

That reference to tacit knowledge and the concurrent rejection of rules isolates another problem that has bothered many of my critics and seemed to provide a basis for charges of subjectivity and irrationality. Some readers have felt that I was trying to make science rest on unanalyzable individual intuitions rather than on logic and law. But that interpretation goes astray in two essential respects. First, if I am talking at all about intuitions, they are not individual. Rather they are the tested and shared possessions of the members of a successful group, and the novice acquires them through training as a part of his preparation for group-membership. Second, they are not in principle unanalyzable. On the contrary, I am currently experimenting with a computer program designed to investigate their properties at an elementary level.

About that program I shall have nothing to say here,[12] but even mention of it should make my most essential point. When I speak of knowledge embedded in shared exemplars, I am not referring to a mode of knowing that is less systematic or less analyzable than knowledge embedded in rules, laws, or criteria of identification. Instead I have in mind a manner of knowing which is miscontrued if reconstructed in terms of rules that are first abstracted from exemplars and thereafter function in their stead. Or, to put the same point differently, when I speak of acquiring from exemplars the ability to recognize a given situation as like some and unlike others that one has seen before, I am not suggesting a process that is not potentially fully explicable in terms of neuro-cerebral mechanism. Instead I am claiming that the explication will not, by its nature, answer the question, "Similar with respect to what?" That question is a request for a rule, in this case for the criteria by which particular situations are grouped into similarity sets, and I am arguing that the temptation to seek criteria (or at least a full set) should be resisted in this case. It is not, however, system but a particular sort of system that I am opposing.

To give that point substance, I must briefly digress. What follows seems obvious to me now, but the constant recourse in my original text to phrases like "the world changes" suggests that it has not always been so. If two people stand at the same place and gaze in the same direction, we must, under pain of solipsism, conclude that they receive closely similar stimuli. (If both could put their eyes at the same place, the stimuli would be identical.) But people do not see stimuli; our knowledge of them is highly theoretical and abstract. Instead they have sensations, and we are under no compulsion to suppose that the sensations of our two viewers are the same. (Sceptics might remember that color blindness was nowhere noticed until John Dalton's description of it in 1794.) On the contrary, much

[12] Some information on this subject can be found in "Second Thoughts."

neural processing takes place between the receipt of a stimulus and the awareness of a sensation. Among the few things that we know about it with assurance are: that very different stimuli can produce the same sensations; that the same stimulus can produce very different sensations; and, finally, that the route from stimulus to sensation is in part conditioned by education. Individuals raised in different societies behave on some occasions as though they saw different things. If we were not tempted to identify stimuli one-to-one with sensations, we might recognize that they actually do so.

Notice now that two groups, the members of which have systematically different sensations on receipt of the same stimuli, do *in some sense* live in different worlds. We posit the existence of stimuli to explain our perceptions of the world, and we posit their immutability to avoid both individual and social solipsism. About neither posit have I the slightest reservation. But our world is populated in the first instance not by stimuli but by the objects of our sensations, and these need not be the same individual to individual or group to group. To the extent, of course, that individuals belong to the same group and thus share education, language, experience, and culture, we have good reason to suppose that their sensations are the same. How else are we to understand the fulness of their communication and the communality of their behavioral responses to their environment? They must see things, process stimuli, in much the same ways. But where the differentiation and specialization of groups begins, we have no similar evidence for the immutability of sensation. Mere parochialism, I suspect, makes us suppose that the route from stimuli to sensation is the same for the members of all groups.

Returning now to exemplars and rules, what I have been trying to suggest, in however preliminary a fashion, is this. One of the fundamental techniques by which the members of a group, whether an entire culture or a specialists' sub-community within it, learn to see the same things when confronted with the same stimuli is by being shown examples of situations that their predecessors in the group have already learned to see as like each other and as different from other sorts of situations. These similar situations may be successive sensory presentations of the same individual—say of mother, who is ultimately recognized on sight as what she is and as different from father or sister. They may be presentations of the members of natural families, say of swans on the one hand and of geese on the other. Or they may, for the members of more specialized groups, be examples of the Newtonian situation, of situations, that is, that are alike in being subject to a version of the symbolic form $f = ma$ and that are different from those situations to which, for example, the law-sketches of optics apply.

Grant for the moment that something of this sort does occur. Ought we say that what has been acquired from exemplars is rules and the ability to apply them? That description is tempting because our seeing a situation as like ones we have encountered before must be the result of neural processing, fully governed by physical and chemical laws. In this sense, once we have learned to do it, recognition of similarity must be as fully systematic as the beating of our hearts. But that very parallel suggests that recognition may also be involuntary, a process over which we have no control. If it is, then we may not properly conceive it as something we manage by applying rules and criteria. To speak of it in those terms implies that we have access to alternatives, that we might, for example, have disobeyed a rule, or misapplied a criterion, or experimented with some other way of seeing.[13] Those, I take it, are just the sorts of things we cannot do.

Or, more precisely, those are things we cannot do until after we have had a sensation, perceived something. Then we do often seek criteria and put them to use. Then we may engage in interpretation, a deliberative process by which we choose among alternatives as we do not in perception itself. Perhaps, for example, something is odd about what we have seen (remember the anomalous playing cards). Turning a corner we see mother

[13] This point might never have needed making if all laws were like Newton's and all rules like the Ten Commandments. In that case the phrase 'breaking a law' would be nonsense, and a rejection of rules would not seem to imply a process not governed by law. Unfortunately, traffic laws and similar products of legislation can be broken, which makes the confusion easy.

entering a downtown store at a time we had thought she was home. Contemplating what we have seen we suddenly exclaim, "That wasn't mother, for she has red hair!" Entering the store we see the woman again and cannot understand how she could have been taken for mother. Or, perhaps we see the tail feathers of a waterfowl feeding from the bottom of a shallow pool. Is it a swan or a goose? We contemplate what we have seen, mentally comparing the tail feathers with those of swans and geese we have seen before. Or, perhaps, being proto-scientists, we simply want to know some general characteristic (the whiteness of swans, for example) of the members of a natural family we can already recognize with ease. Again, we contemplate what we have previously perceived, searching for what the members of the given family have in common.

These are all deliberative processes, and in them we do seek and deploy criteria and rules. We try, that is, to interpret sensations already at hand, to analyze what is for us the given. However we do that, the processes involved must ultimately be neural, and they are therefore governed by the same *physico-chemical* laws that govern perception on the one hand and the beating of our hearts on the other. But the fact that the system obeys the same laws in all three cases provides no reason to suppose that our neural apparatus is programmed to operate the same way in interpretation as in perception or in either as in the beating of our hearts. What I have been opposing in this book is therefore the attempt, traditional since Descartes but not before, to analyze perception as an interpretive process, as an unconscious version of what we do after we have perceived.

What makes the integrity of perception worth emphasizing is, of course, that so much past experience is embodied in the neural apparatus that transforms stimuli to sensations. An appropriately programmed perceptual mechanism has survival value. To say that the members of different groups may have different perceptions when confronted with the same stimuli is not to imply that they may have just any perceptions at all. In many environments a group that could not tell wolves from dogs could not endure. Nor would a group of nuclear physicists today survive as scientists if unable to recognize the tracks of alpha particles and electrons. It is just because so very few ways of seeing will do that the ones that have withstood the tests of group use are worth transmitting from generation to generation. Equally, it is because they have been selected for their success over historic time that we must speak of the experience and knowledge of nature embedded in the stimulus-to-sensation route.

Perhaps 'knowledge' is the wrong word, but there are reasons for employing it. What is built into the neural process that transforms stimuli to sensations has the following characteristics: it has been transmitted through education; it has, by trial, been found more effective than its historical competitors in a group's current environment; and, finally, it is subject to change both through further education and through the discovery of misfits with the environment. Those are characteristics of knowledge, and they explain why I use the term. But it is strange usage, for one other characteristic is missing. We have no direct access to what it is we know, no rules or generalizations with which to express this knowledge. Rules which could supply that access would refer to stimuli not sensations, and stimuli we can know only through elaborate theory. In its absence, the knowledge embedded in the stimulus-to-sensation route remains tacit.

Though it is obviously preliminary and need not be correct in all details, what has just been said about sensation is meant literally. At the very least it is a hypothesis about vision which should be subject to experimental investigation though probably not to direct check. But talk like this of seeing and sensation here also serves metaphorical functions as it does in the body of the book. We do not *see* electrons, but rather their tracks or else bubbles of vapor in a cloud chamber. We do not *see* electric currents at all, but rather the needle of an ammeter or galvanometer. Yet in the preceding pages, particularly in Section X, I have repeatedly acted as though we did perceive theoretical entities like currents, electrons, and fields, as though we learned to do so from examination of exemplars, and as though in these cases too it would be wrong to replace talk of seeing with talk of criteria and interpretation. The metaphor that transfers 'seeing' to contexts like these is scarcely a sufficient basis for such claims. In the long run it will need to be eliminated in favor of a more literal mode of discourse.

The computer program referred to above begins to suggest ways in which that may be done, but neither available space nor the extent of my present understanding permits my eliminating the metaphor here.[14] Instead I shall try briefly to bulwark it. Seeing water droplets or a needle against a numerical scale is a primitive perceptual experience for the man unacquainted with cloud chambers and ammeters. It thus requires contemplation, analysis, and interpretation (or else the intervention of external authority) before conclusions can be reached about electrons or currents. But the position of the man who has learned about these instruments and had much exemplary experience with them is very different, and there are corresponding differences in the way he processes the stimuli that reach him from them. Regarding the vapor in his breath on a cold winter afternoon, his sensation may be the same as that of a layman, but viewing a cloud chamber he sees (here literally) not droplets but the tracks of electrons, alpha particles, and so on. Those tracks are, if you will, criteria that he interprets as indices of the presence of the corresponding particles, but that route is both shorter and different from the one taken by the man who interprets droplets.

Or consider the scientist inspecting an ammeter to determine the number against which the needle has settled. His sensation probably is the same as the layman's, particularly if the latter has

[14] For readers of "Second Thoughts" the following cryptic remarks may be leading. The possibility of immediate recognition of the members of natural families depends upon the existence, after neural processing, of empty perceptual space between the families to be discriminated. If, for example, there were a perceived continuum of waterfowl ranging from geese to swans, we should be compelled to introduce a specific criterion for distinguishing them. A similar point can be made for unobservable entities. If a physical theory admits the existence of nothing else like an electric current, then a small number of criteria, which may vary considerably from case to case, will suffice to identify currents even though there is no set of rules that specifies the necessary and sufficient conditions for the identification. That point suggests a plausible corollary which may be more important. Given a set of necessary and sufficient conditions for identifying a theoretical entity, that entity can be eliminated from the ontology of a theory by substitution. In the absence of such rules, however, these entities are not eliminable; the theory then demands their existence.

read other sorts of meters before. But he has seen the meter (again often literally) in the context of the entire circuit, and he knows something about its internal structure. For him the needle's position is a criterion, but only of *the value* of the current. To interpret it he need determine only on which scale the meter is to be read. For the layman, on the other hand, the needle's position is not a criterion of anything except itself. To interpret it, he must examine the whole layout of wires, internal and external, experiment with batteries and magnets, and so on. In the metaphorical no less than in the literal use of 'seeing,' interpretation begins where perception ends. The two processes are not the same, and what perception leaves for interpretation to complete depends drastically on the nature and amount of prior experience and training.

5. *Exemplars, Incommensurability, and Revolutions*

What has just been said provides a basis for clarifying one more aspect of the book: my remarks on incommensurability and its consequences for scientists debating the choice between successive theories.[15] In Sections X and XII I have argued that the parties to such debates inevitably see differently certain of the experimental or observational situations to which both have recourse. Since the vocabularies in which they discuss such situations consist, however, predominantly of the same terms, they must be attaching some of those terms to nature differently, and their communication is inevitably only partial. As a result, the superiority of one theory to another is something that cannot be proved in the debate. Instead, I have insisted, each party must try, by persuasion, to convert the other. Only philosophers have seriously misconstrued the intent of these parts of my argument. A number of them, however, have reported that I believe the following:[16] the proponents of incommensurable theories

[15] The points that follow are dealt with in more detail in Secs. v and vi of "Reflections."

[16] See the works cited in note 9, above, and also the essay by Stephen Toulmin in *Growth of Knowledge*.

cannot communicate with each other at all; as a result, in a debate over theory-choice there can be no recourse to *good* reasons; instead theory must be chosen for reasons that are ultimately personal and subjective; some sort of mystical apperception is responsible for the decision actually reached. More than any other parts of the book, the passages on which these misconstructions rest have been responsible for charges of irrationality.

Consider first my remarks on proof. The point I have been trying to make is a simple one, long familiar in philosophy of science. Debates over theory-choice cannot be cast in a form that fully resembles logical or mathematical proof. In the latter, premises and rules of inference are stipulated from the start. If there is disagreement about conclusions, the parties to the ensuing debate can retrace their steps one by one, checking each against prior stipulation. At the end of that process one or the other must concede that he has made a mistake, violated a previously accepted rule. After that concession he has no recourse, and his opponent's proof is then compelling. Only if the two discover instead that they differ about the meaning or application of stipulated rules, that their prior agreement provides no sufficient basis for proof, does the debate continue in the form it inevitably takes during scientific revolutions. That debate is about premises, and its recourse is to persuasion as a prelude to the possibility of proof.

Nothing about that relatively familiar thesis implies either that there are no good reasons for being persuaded or that those reasons are not ultimately decisive for the group. Nor does it even imply that the reasons for choice are different from those usually listed by philosophers of science: accuracy, simplicity, fruitfulness, and the like. What it should suggest, however, is that such reasons function as values and that they can thus be differently applied, individually and collectively, by men who concur in honoring them. If two men disagree, for example, about the relative fruitfulness of their theories, or if they agree about that but disagree about the relative importance of fruitfulness and, say, scope in reaching a choice, neither can be convicted of a mistake. Nor is either being unscientific. There is no neutral algorithm for theory-choice, no systematic decision procedure which, properly applied, must lead each individual in the group to the same decision. In this sense it is the community of specialists rather than its individual members that makes the effective decision. To understand why science develops as it does, one need not unravel the details of biography and personality that lead each individual to a particular choice, though that topic has vast fascination. What one must understand, however, is the manner in which a particular set of shared values interacts with the particular experiences shared by a community of specialists to ensure that most members of the group will ultimately find one set of arguments rather than another decisive.

That process is persuasion, but it presents a deeper problem. Two men who perceive the same situation differently but nevertheless employ the same vocabulary in its discussion must be using words differently. They speak, that is, from what I have called incommensurable viewpoints. How can they even hope to talk together much less to be persuasive. Even a preliminary answer to that question demands further specification of the nature of the difficulty. I suppose that, at least in part, it takes the following form.

The practice of normal science depends on the ability, acquired from exemplars, to group objects and situations into similarity sets which are primitive in the sense that the grouping is done without an answer to the question, "Similar with respect to what?" One central aspect of any revolution is, then, that some of the similarity relations change. Objects that were grouped in the same set before are grouped in different ones afterward and vice versa. Think of the sun, moon, Mars, and earth before and after Copernicus; of free fall, pendular, and planetary motion before and after Galileo; or of salts, alloys, and a sulpuhur–iron filing mix before and after Dalton. Since most objects within even the altered sets continue to be grouped together, the names of the sets are usually preserved. Nevertheless, the transfer of a subset is ordinarily part of a critical change in the network of interrelations among them. Transferring the metals from the set of compounds to the set of elements played an essential role in the emergence of a new theory of combustion, of acidity, and of physical and chemical combination. In short order those changes had spread through all of chemistry. Not surprisingly, therefore, when such redistributions occur, two men whose discourse had previously proceeded with apparently full understanding may suddenly find themselves responding to the same stimulus with incompatible descriptions and generalizations. Those difficulties will not be felt in all areas of even their scientific discourse, but they will arise and will then cluster most densely about the phenomena upon which the choice of theory most centrally depends.

Such problems, though they first become evident in communication, are not merely linguistic, and they cannot be resolved simply by stipulating the definitions of troublesome terms. Because the words about which difficulties cluster have been learned in part from direct application to exemplars, the participants in a communication breakdown cannot say, "I use the word 'element' (or 'mixture,' or 'planet,' or 'unconstrained motion') in ways determined by the following criteria." They cannot, that is, resort to a neutral language which both use in the same way and which is adequate to the statement of both their theories or even of both those theories' empirical consequences. Part of the difference is prior to the application of the languages in which it is nevertheless reflected.

The men who experience such communication breakdowns must, however, have some recourse. The stimuli that impinge upon them are the same. So is their general neural apparatus, however differently programmed. Furthermore, except in a small, if all-important, area of experience even their neural programming must be very nearly the same, for they share a history, except the immediate past. As a result, both their everyday and most of their scientific world and language are shared. Given that much in common, they should be able to find out a great deal about how they differ. The techniques required are not, however, either straightforward, or comfortable, or parts of the scientist's normal arsenal. Scientists rarely recognize them for quite what they are, and they seldom use them for longer than is required to induce conversion or convince themselves that it will not be obtained.

Briefly put, what the participants in a communication breakdown can do is recognize each other as members of different language communities and then become translators.[17] Taking the differences between their own intra- and inter-group discourse as itself a subject for study, they can first attempt to discover the terms and locutions that, used unproblematically within each community, are nevertheless foci of trouble for inter-group discussions. (Locutions that present no such difficulties may be homophonically translated.) Having isolated such areas of difficulty in scientific communication, they can next resort to their shared everyday vocabularies in an effort further to elucidate their troubles. Each may, that is, try to discover what the other would see and say when presented with a stimulus to which his own verbal response would be different. If they can sufficiently refrain from explaining anomalous behavior as the consequence of mere error or madness, they may in time become very good predictors of each other's behavior. Each will have learned to translate the other's theory and its consequences into his own language and simultaneously to describe in his language the world to which that theory applies. That is what the historian of science regularly does (or should) when dealing with out-of-date scientific theories.

Since translation, if pursued, allows the participants in a communication breakdown to experience vicariously something of the merits and defects of each other's points of view, it is a potent tool both for persuasion and for conversion. But even persuasion need not succeed, and, if it does, it need not be

[17] The already classic source for most of the relevant aspects of translation is W. V. O. Quine, *Word and Object* (Cambridge, Mass., and New York, 1960), chaps. i and ii. But Quine seems to assume that two men receiving the same stimulus must have the same sensation and therefore has little to say about the extent to which a translator must be able to *describe* the world to which the language being translated applies. For the latter point see, E. A. Nida, "Linguistics and Ethnology in Translation Problems," in Del Hymes (ed.), *Language and Culture in Society* (New York, 1964), pp. 90–97.

accompanied or followed by conversion. The two experiences are not the same, an important distinction that I have only recently fully recognized.

To persuade someone is, I take it, to convince him that one's own view is superior and ought therefore supplant his own. That much is occasionally achieved without recourse to anything like translation. In its absence many of the explanations and problem-statements endorsed by the members of one scientific group will be opaque to the other. But each language community can usually produce from the start a few concrete research results that, though describable in sentences understood in the same way by both groups, cannot yet be accounted for by the other community in its own terms. If the new viewpoint endures for a time and continues to be fruitful, the research results verbalizable in this way are likely to grow in number. For some men such results alone will be decisive. They can say: I don't know how the proponents of the new view succeed, but I must learn; whatever they are doing, it is clearly right. That reaction comes particularly easily to men just entering the profession, for they have not yet acquired the special vocabularies and commitments of either group.

Arguments statable in the vocabulary that both groups use in the same way are not, however, usually decisive, at least not until a very late stage in the evolution of the opposing views. Among those already admitted to the profession, few will be persuaded without some recourse to the more extended comparisons permitted by translation. Though the price is often sentences of great length and complexity (think of the Proust-Berthollet controversy conducted without recourse to the term 'element'), many additional research results can be *translated* from one community's language into the other's. As translation proceeds, furthermore, some members of each community may also begin vicariously to understand how a statement previously opaque could seem an explanation to members of the opposing group. The availability of techniques like these does not, of course, guarantee persuasion. For most people translation is a threatening process, and it is entirely foreign to normal science.

Counter-arguments are, in any case, always available, and no rules prescribe how the balance must be struck. Nevertheless, as argument piles on argument and as challenge after challenge is successfully met, only blind stubbornness can at the end account for continued resistance.

That being the case, a second aspect of translation, long familiar to both historians and linguists, becomes crucially important. To translate a theory or worldview into one's own language is not to make it one's own. For that one must go native, discover that one is thinking and working in, not simply translating out of, a language that was previously foreign. That transition is not, however, one that an individual may make or refrain from making by deliberation and choice, however good his reasons for wishing to do so. Instead, at some point in the process of learning to translate, he finds that the transition has occurred, that he has slipped into the new language without a decision having been made. Or else, like many of those who first encountered, say, relativity or quantum mechanics in their middle years, he finds himself fully persuaded of the new view but nevertheless unable to internalize it and be at home in the world it helps to shape. Intellectually such a man has made his choice, but the conversion required if it is to be effective eludes him. He may use the new theory nonetheless, but he will do so as a foreigner in a foreign environment, an alternative available to him only because there are natives already there. His work is parasitic on theirs, for he lacks the constellation of mental sets which future members of the community will acquire through education.

The conversion experience that I have likened to a gestalt switch remains, therefore, at the heart of the revolutionary process. Good reasons for choice provide motives for conversion and a climate in which it is more likely to occur. Translation may, in addition, provide points of entry for the neural reprogramming that, however inscrutable at this time, must underlie conversion. But neither good reasons nor translation constitute conversion, and it is that process we must explicate in order to understand an essential sort of scientific change.

6. *Revolutions and Relativism*

One consequence of the position just outlined has particularly bothered a number of my critics.[18] They find my viewpoint relativistic, particularly as it is developed in the last section of this book. My remarks about translation highlight the reasons for the charge. The proponents of different theories are like the members of different language-culture communities. Recognizing the parallelism suggests that in some sense both groups may be right. Applied to culture and its development that position is relativistic.

But applied to science it may not be, and it is in any case far from *mere* relativism in a respect that its critics have failed to see. Taken as a group or in groups, practitioners of the developed sciences are, I have argued, fundamentally puzzle-solvers. Though the values that they deploy at times of theory-choice derive from other aspects of their work as well, the demonstrated ability to set up and to solve puzzles presented by nature is, in case of value conflict, the dominant criterion for most members of a scientific group. Like any other value, puzzle-solving ability proves equivocal in application. Two men who share it may nevertheless differ in the judgments they draw from its use. But the behavior of a community which makes it preeminent will be very different from that of one which does not. In the sciences, I believe, the high value accorded to puzzle-solving ability has the following consequences.

Imagine an evolutionary tree representing the development of the modern scientific specialties from their common origins in, say, primitive natural philosophy and the crafts. A line drawn up that tree, never doubling back, from the trunk to the tip of some branch would trace a succession of theories related by descent. Considering any two such theories, chosen from points not too near their origin, it should be easy to design a list of criteria that would enable an uncommitted observer to distinguish the earlier from the more recent theory time after time. Among

[18] Shapere, "Structure of Scientific Revolutions," and Popper in *Growth of Knowledge*.

the most useful would be: accuracy of prediction, particularly of quantitative prediction; the balance between esoteric and everyday subject matter; and the number of different problems solved. Less useful for this purpose, though also important determinants of scientific life, would be such values as simplicity, scope, and compatibility with other specialties. Those lists are not yet the ones required, but I have no doubt that they can be completed. If they can, then scientific development is, like biological, a unidirectional and irreversible process. Later scientific theories are better than earlier ones for solving puzzles in the often quite different environments to which they are applied. That is not a relativist's position, and it displays the sense in which I am a convinced believer in scientific progress.

Compared with the notion of progress most prevalent among both philosophers of science and laymen, however, this position lacks an essential element. A scientific theory is usually felt to be better than its predecessors not only in the sense that it is a better instrument for discovering and solving puzzles but also because it is somehow a better representation of what nature is really like. One often hears that successive theories grow ever closer to, or approximate more and more closely to, the truth. Apparently generalizations like that refer not to the puzzle-solutions and the concrete predictions derived from a theory but rather to its ontology, to the match, that is, between the entities with which the theory populates nature and what is "really there."

Perhaps there is some other way of salvaging the notion of 'truth' for application to whole theories, but this one will not do. There is, I think, no theory-independent way to reconstruct phrases like 'really there'; the notion of a match between the ontology of a theory and its "real" counterpart in nature now seems to me illusive in principle. Besides, as a historian, I am impressed with the implausability of the view. I do not doubt, for example, that Newton's mechanics improves on Aristotle's and that Einstein's improves on Newton's as instruments for puzzle-solving. But I can see in their succession no coherent direction of ontological development. On the contrary, in some important respects, though by no means in all, Einstein's general theory of relativity is closer to Aristotle's than either of them is to Newton's. Though the temptation to describe that position as relativistic is understandable, the description seems to me wrong. Conversely, if the position be relativism, I cannot see that the relativist loses anything needed to account for the nature and development of the sciences.

7. *The Nature of Science*

I conclude with a brief discussion of two recurrent reactions to my original text, the first critical, the second favorable, and neither, I think, quite right. Though the two relate neither to what has been said so far nor to each other, both have been sufficiently prevalent to demand at least some response.

A few readers of my original text have noticed that I repeatedly pass back and forth between the descriptive and the normative modes, a transition particularly marked in occasional passages that open with, "But that is not what scientists do," and close by claiming that scientists ought not do so. Some critics claim that I am confusing description with prescription, violating the time-honored philosophical theorem: 'Is' cannot imply 'ought.'[19]

That theorem has, in practice, become a tag, and it is no longer everywhere honored. A number of contemporary philosophers have discovered important contexts in which the normative and the descriptive are inextricably mixed.[20] 'Is' and 'ought' are by no means always so separate as they have seemed. But no recourse to the subtleties of contemporary linguistic philosophy is needed to unravel what has seemed confused about this aspect of my position. The preceding pages present a viewpoint or theory about the nature of science, and, like other philosophies of science, the theory has consequences for the way in which scientists should behave if their enterprise is to succeed. Though

[19] For one of many examples, see P. K. Feyerabend's essay in *Growth of Knowledge*.

[20] Stanley Cavell, *Must We Mean What We Say?* (New York, 1969), chap. i.

it need not be right, any more than any other theory, it provides a legitimate basis for reiterated 'oughts' and 'shoulds.' Conversely, one set of reasons for taking the theory seriously is that scientists, whose methods have been developed and selected for their success, do in fact behave as the theory says they should. My descriptive generalizations are evidence for the theory precisely because they can also be derived from it, whereas on other views of the nature of science they constitute anomalous behavior. The circularity of that argument is not, I think, vicious. The consequences of the viewpoint being discussed are not exhausted by the observations upon which it rested at the start. Even before this book was first published, I had found parts of the theory it presents a useful tool for the exploration of scientific behavior and development. Comparison of this postscript with the pages of the original may suggest that it has continued to play that role. No merely circular point of view can provide such guidance.

To one last reaction to this book, my answer must be of a different sort. A number of those who have taken pleasure from it have done so less because it illuminates science than because they read its main theses as applicable to many other fields as well. I see what they mean and would not like to discourage their attempts to extend the position, but their reaction has nevertheless puzzled me. To the extent that the book portrays scientific development as a succession of tradition-bound periods punctuated by non-cumulative breaks, its theses are undoubtedly of wide applicability. But they should be, for they are borrowed from other fields. Historians of literature, of music, of the arts, of political development, and of many other human activities have long described their subjects in the same way. Periodization in terms of revolutionary breaks in style, taste, and institutional structure have been among their standard tools. If I have been original with respect to concepts like these, it has mainly been by applying them to the sciences, fields which had been widely thought to develop in a different way. Conceivably the notion of a paradigm as a concrete achievement, an exemplar, is a second contribution. I suspect, for example, that some of the notorious difficulties surrounding the notion of style in the arts may vanish if paintings can be seen to be modeled on one another rather than produced in conformity to some abstracted canons of style.[21]

This book, however, was intended also to make another sort of point, one that has been less clearly visible to many of its readers. Though scientific development may resemble that in other fields more closely than has often been supposed, it is also strikingly different. To say, for example, that the sciences, at least after a certain point in their development, progress in a way that other fields do not, cannot have been all wrong, whatever progress itself may be. One of the objects of the book was to examine such differences and begin accounting for them.

Consider, for example, the reiterated emphasis, above, on the absence or, as I should now say, on the relative scarcity of competing schools in the developed sciences. Or remember my remarks about the extent to which the members of a given scientific community provide the only audience and the only judges of that community's work. Or think again about the special nature of scientific education, about puzzle-solving as a goal, and about the value system which the scientific group deploys in periods of crisis and decision. The book isolates other features of the same sort, none necessarily unique to science but in conjunction setting the activity apart.

About all these features of science there is a great deal more to be learned. Having opened this postscript by emphasizing the need to study the community structure of science, I shall close by underscoring the need for similar and, above all, for comparative study of the corresponding communities in other fields. How does one elect and how is one elected to membership in a particular community, scientific or not? What is the process and what are the stages of socialization to the group? What does the group collectively see as its goals; what deviations, individual or collective, will it tolerate; and how does it control the impermissible aberration? A fuller understanding of science will de-

[21] For this point as well as a more extended discussion of what is special about the sciences, see T. S. Kuhn, "Comment [on the Relations of Science and Art]," *Comparative Studies in Philosophy and History*, XI (1969), 403–12.

pend on answers to other sorts of questions as well, but there is no area in which more work is so badly needed. Scientific knowledge, like language, is intrinsically the common property of a group or else nothing at all. To understand it we shall need to know the special characteristics of the groups that create and use it.